Microbial Symbioses

Series Editor
Françoise Gaill

Microbial Symbioses

Sébastien Duperron

First published 2017 in Great Britain and the United States by ISTE Press Ltd and Elsevier Ltd

ISTE Press Ltd
27-37 St George's Road
London SW19 4EU
UK

www.iste.co.uk

Elsevier Ltd
The Boulevard, Langford Lane
Kidlington, Oxford, OX5 1GB
UK

www.elsevier.com

British Library Cataloguing-in-Publication Data
A CIP record for this book is available from the British Library
Library of Congress Cataloging in Publication Data
A catalog record for this book is available from the Library of Congress
ISBN 978-1-78548-220-5

Printed and bound in the UK and US

Contents

Acknowledgments

This book could not have been written without the support of Françoise Gaill (CNRS), coordinator of ISTE's Ecological Science collection, and of the editorial team. It has benefited greatly from the comments of Yves Desdevises (UPMC), Paola Furla (University of Nice), Olivier Gros (University of the French West Indies and Guiana), Natacha Kremer (CNRS), Bérénice Piquet (UPMC) and Claire Tirard (UPMC). I would also like to thank David Sillam-Dusses (University of Paris 13), Benjamin Loppin (CNRS) and Margaret McFall-Ngai (University of Hawaii), who permitted me to use photos of the models they are working on. The histological cross-section illustrations come from UPMC's remarkable Biology teaching collection, which has been maintained and supplemented by the work of many people over many years. Finally, I would like to thank the *Institut Universitaire de France* for its support.

Sébastien DUPERRON

Introduction

Symbiosis. In common usage, it is a portmanteau that denotes any type of interaction perceived as beneficial in any field, whether it be management, construction, waste management, or even business mergers and acquisitions. A word that originated in the vocabulary of biological sciences has come to be used by both managers and ordinary people. Symbiosis has very positive connotations in our societies, perhaps because it offers a soothing alternative to the individualism and competition that are often rife within them, described by Thomas Hobbes as "the war of all against all". For biologists, symbiosis denotes certain types of interaction between species. To study it is to enter an immense multidisciplinary field whose precise scope is hard to define. The concept of symbiosis contributes to a holistic vision of living things that is markedly different to the reductionist approaches often used by biologists. This may be why symbiosis has struggled to find its rightful place in life science research and teaching remains too often limited to describing a few anecdotal examples of interactions.

But times change. Discoveries made in recent years, facilitated by the development of new high-throughput techniques in life sciences, show that plants and animals evolve from their origins in a principally microbial world, with which they interact closely. These interactions, these microbial symbioses, represent a significantly expanding field of study. Researchers are discovering new symbioses involving new agents and new functions in all types of habitats, from the ocean floors to the poles and the highest peaks. The human body itself appears increasingly to be a complex ecosystem, within which bacteria play the most important roles. Our planet's microscopic agents have been found to have shaped not only our environment but also our bodies, right down to our behavior, ever since complex cells with a nucleus first appeared. Microbial symbiosis plays a central role in this ongoing paradigm shift. It has implications for the very concept of the individual, as well as for traditional representations of the evolution of species. The study of symbioses, particularly those involving microorganisms, is now a dynamic discipline, transversal in its aims and concerns and highly promising for applications in diverse fields.

This book provides an overview of current research in the field of microbial symbioses. The first chapter introduces the concept of symbiosis, which appeared in the second half of the 19th Century and has evolved up to the present day. The term and the tools accessible to the researcher, a factor that often restricts the questions that can be answered, have evolved in parallel. Chapters 2 and 3 discuss the many roles of symbiosis that have been discovered, separating those directly related to nutrition, which are often the best documented, from those that are more unexpected. A comparative study of the various examples reveals some general rules of how symbiosis works: this takes the place of a study framework and is the subject of Chapter 4. The workings of symbiosis, a game-changer when it comes to the relationships between organisms, are both the result of the partners' prior evolution and the impetus behind the future evolution of the entity they form. Chapter 5 discusses symbiosis as a source of novelty in evolution. Chapter 6 outlines its important role in the current functioning as well as in the emergence of the biosphere that we know today. The most recent discoveries have even made symbiosis a source of inspiration for fundamental and applied research, leading researchers to explore particularly promising avenues in many areas, as discussed in Chapter 7. At the end of this journey, which provides a deliberately "symbiocentric" perspective on biology, we will assess the ways in which the exploration of microbial symbioses has changed our understanding of Life on Earth, and what the future may hold for this discipline.

1

The Concept of Symbiosis, from Past to Present

1.1. A brief history

Since ancient times, authors have cited examples of mutual aid between species. Herodotus describes the relationship between the Nile crocodile and the Egyptian plover, a small bird that nests on riverbanks. The bird cleans the crocodile's teeth, removing fragments of food and parasites from between them. The crocodile, mouth open wide, does not attempt to devour it. However, the concept of symbiosis appeared only in the second half of the 19th Century, 20 years after the publication of Darwin's "The Origin of Species" in 1859 [DAR 59]. A detailed history of the concept of symbiosis is beyond the scope of this book, but a small number of authors, ideas and controversies illustrating its emergence and evolution will be outlined, providing a historical context for the recent discoveries discussed later on. Interested readers will find a more exhaustive overview in a number of works on the subject [BUC 65, SAP 94].

1.1.1. *Emergence of the concept of symbiosis*

In 1878, when Heinrich Anton de Bary (1831–1888) defined symbiosis as the living together of organisms belonging to different species, scientists had already been studying the interactions between organisms for many decades. The word "parasite" (from *para*, "alongside", and *sitos*, "corn, bread"), for example, had been used since the Middle Ages. During the 1860s, the lichens that Anton de Bary studied were still considered to be plants. But, in 1866, he suggested that some lichens could be associations between an alga and a fungus. In 1867, the Swiss botanist Simon Schwendener (1829–1919) was the first to suggest that all lichens have a dual nature, associating an alga and a fungus in a relationship akin to slavery. This idea was unpopular with systematists, since it destroyed the foundations of a classification system that related all organisms to a species. The success of algae isolation quickly confirmed Schwendener's hypothesis, but his interpretation of the relationship as slavery-based or parasitic was rejected [HON 00]. The vocabulary used to describe relationships between organisms broadened with the turn of the 1870s. In 1873, the Belgian zoologist Pierre-Joseph van Beneden (1809–1894) introduced the concepts of mutualism and commensalism to biology, formalizing them in an 1875 work entitled "*Les commensaux et les parasites*" ("Animal Parasites and Messmates") [VAN 75]. This work distinguished parasitism, commensalism and mutualism (Figure 1.1). In parasitism, one partner survives to the detriment of the other. Today, the parasite is said to have a negative effect on the fitness of its partner. Commensalism (from the Latin *cum*, "with", and *mensa*, "table") is a relationship in which the commensal benefits from its association with its host (e.g. obtaining food), without damaging it. Finally, mutualism is a reciprocal relationship in which each partner draws a net benefit, i.e. one that exceeds any cost of the relationship.

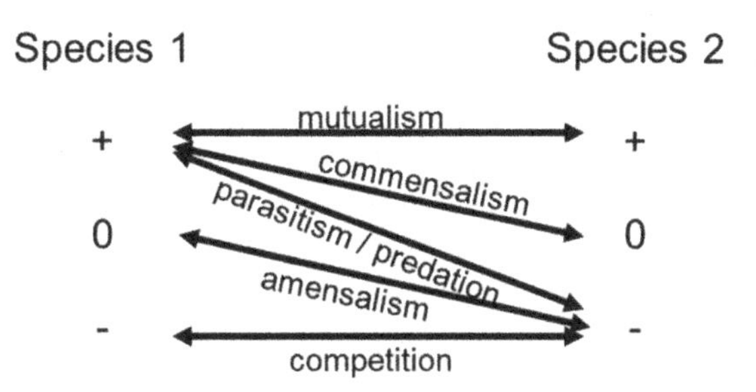

Figure 1.1. *Classification of inter-species relationships according to their impact on each of the two species. An example of amensalism, a relationship that damages one partner without benefiting the other, is the trampling of fragile plant species by tourists on a dune*

Van Beneden highlighted that there is actually a continuous gradation between these different types of relationship, and Anton de Bary believed that the nature of the relationship between algae and fungi in lichens is not necessarily unique. A more neutral term was required, based on the coexistence itself rather than on any consideration of role and conceding that the situation can result in the partners no longer being able to survive without one another. In 1877, Albert Bernhard Frank (1839–1900) proposed the term *symbiotismus* to describe the relationship between partners in lichens, which, he demonstrated, was nothing like parasitism, since the "parasite" is involved in its host's nutrition [FRA 77]. However, Anton de Bary, at the University of Strasbourg (then German) in the 1870s, would be credited with coining this term, following a lecture entitled "*De la symbiose*" ("On Symbiosis"), which he gave in Cassel in 1878. Some of his writings refer to Frank's article, which introduced the term. Anton de Bary's definition of symbiosis included diverse relationships, from parasitism to mutualism, as well as rather stretched inter-species relationships. He even introduced the idea that symbiosis can lead to non-pathological morphological variations, suggesting that it is therefore a source of evolutionary novelty. However, he did not support the idea that it could produce major novelties.

During the same period, various works on the interaction between algae and animals were published. From the 1840s onwards, many researchers observed the green color of some animals and suggested that they contained chlorophyll, the green pigment that intercepts light energy in algae and plants. It was often supposed to be produced by the animal itself and was therefore considered "animal chlorophyll". Karl Brandt (1854–1931) was among the first to show that this color is actually due to the presence of phototrophic microorganisms. The microorganisms associated with the freshwater hydra were named *Zoochlorella* (now identified as green micro-algae), and those found in sea anemones and radiolarians were named *Zooxanthella* in 1881 (the latter now identified as dinoflagellates) [BRA 81]. His work showed that these "algae" could be separated and made to live outside their host. Cienkowski had been publishing similar observations since 1871, but his work was in Hungarian and was therefore not well known. Also in 1881, the naturalist Joseph Leidy (1823–1891), famous for being a dinosaur fossil hunter, identified flagellates in the digestive tube of termites and realized that the interaction was not parasitic [WIE 00]. During the 1890s, Pierre Mazé (1868–1947) isolated bacteria from the nodules of leguminous plants at the Pasteur Institute and demonstrated that they contribute to nitrogen acquisition. In 1885, Albert Frank proposed the term "mycorrhiza" to denote fungal filaments associated with plant roots, which he argued were symbionts [FRA 04]. Research into symbioses therefore developed very quickly. Symbioses involving bacteria associated with animals were recognized a little later, at the turn of the 20th Century. This delay was mainly due to the difficulty of distinguishing bacteria, which are smaller than algae and fungi, and of cultivating the bacterial symbionts from animal hosts to confirm their status as distinct organisms. The latter is still a problem, as we will see in the next sections. It took many years of in-depth study from 1884 onwards for Blochmann, for example, to propose that the rods he had observed in the tissues of some ants could be bacteria, and it was only in 1906 that Mercier's work definitively dispelled all doubt as to their nature.

The concept of symbiosis thus reached a point when a lot of work was being done on the interactions between organisms. As soon as the term was defined, discoveries began to pile up and the concept rapidly increased in substance and popularity. Earlier, in "The Origin of Species", published in 1859, Darwin had mentioned a number of positive interactions, such as the pollination of flowering plants by insects, but this discussion was shaped by the idea that the struggle for existence depended on conflict and competition, rather than on cooperation. The idea that symbiotic interactions could evolve conflicts with this principle, particularly in the case of reciprocally beneficial relationships. This is probably why the term "symbiosis" was usually used to describe mutualist associations, diverging from Anton de Bary's broader definition. There is some ambiguity in the meaning of the word "symbiosis", which, far from being the result of an inadequate initial definition, is due to the context in which it has been preferentially used.

1.1.2. *Symbiosis at the beginning of the 20th Century*

Confirmation of the bacterial nature of the symbionts of many insects sparked the study of bacterial symbioses, and more and more discoveries were made. Another major undertaking was the investigation of the symbiotic or non-symbiotic origins of the organelles of the eukaryotic cell. In 1883 and 1885, Andreas Schimper (1856–1901) suggested that plants might originate in the union of one colorless organism and another colored by chlorophyll. In 1893, Shôsaburo Watasé suggested a similar hypothesis for other organelles, such as the nucleus, which he thought might also be an organism. The theory of symbiogenesis was developed in Russia, under the principal influence of Konstantin

Mereschkowski (1855–1921), who proposed the term in 1910, and Andreï Famintsyne (1835–1918). The idea was that new complex organisms are created through the combination of organisms that enter into symbiosis. This novel idea did not have a great impact at the time, partly due to the lack of convincing proof. In France, Paul Portier's (1866–1962) 1918 work "*Les Symbiotes*" ("Symbionts") proposed symbiosis as a universal mechanism of living things, supporting the idea that mitochondria were bacteria. His theory states that bacteria are the only individualized organisms and that all others come from symbiotic association between them. Although this view may seem modern, many of Portier's interpretations and conclusions were highly controversial, discredited and even ridiculed. For example, his experiments relied on cultivating bacteria from animal organs, which Pasteurian microbiologists immediately considered contamination. He is nonetheless credited with stimulating interest in the subject. Ivan Wallin (1883–1969) made the same hypothesis about the bacterial nature of mitochondria, this time using experimental markings, independently, from 1922 onwards. He was unable to cultivate mitochondria outside of eukaryotic cells and therefore also struggled to convince.

On the edge of these controversies, Paul Buchner (1886–1978), a major player in this period and a specialist in symbionts, especially in insects, published several works on symbiosis from 1921 onwards. The most famous, which appeared in 1953, is a remarkable book entitled "*Endosymbiose der Tiere mit plflanzlichen Mikroorganismen*" (Endosymbioses of Animals with Plant Microorganisms). The fourth edition, expanded and translated into English in 1965, provided a complete overview of the knowledge available at that time, including the history of individual discoveries in various groups [BUC 65]. Although he supported the idea of phototrophic symbiosis, Buchner distanced himself from Portier and from the idea that cells might originate in symbiosis. Instead, he focused on gathering direct proof of its importance in nature. His report is often considered to have founded the more rigorous modern study of symbiosis. In fact, it was not until the 1960s and 1970s that Lynn Margulis (1938–2011) approached bold ideas with the scientific rigor made possible by the improvement of techniques, solidly demonstrating the endosymbiotic origins of the eukaryotic cell [MAR 71].

1.2. Defining symbiosis

1.2.1. *Classical definitions and difficulties*

The word "symbiosis" comes from the Greek "*sumbioûn*", which literally means "to live together". Anton de Bary defines symbiosis as the living together (or association) of dissimilar organisms (i.e. belonging to different species). We have seen how this initially very broad definition had evolved with past discoveries, briefly being confused with mutualism (Figure 1.1). In French, the word "symbiosis" retains this connotation and is still often used (and taught) as a synonym of mutualism, even of obligate mutualism. The Larousse Dictionary, for example, defines it as the constant, obligatory and specific association between two organisms that cannot live without one another, each of them benefiting from the association. The Cambridge Dictionary defines symbiosis as a relationship between two types of animals or plants in which each provides the other with the conditions necessary for its continued existence. The English meaning is broader and incorporates forms of interaction that do not necessarily benefit each partner, such as commensalism, in which the association is beneficial for one and neutral for the other, and parasitism, in which the benefit for one is obtained to the detriment of the other, and even mutually negative relationships. How, therefore, do we define the concept? This is not a new problem: in 1897, for example, the lichens specialist Albert Schneider suggested that the definition should be limited to associations that lead to the loss or gain of nutrients, and alternative definition attempts have been proposed in a number of publications since.

The concept of benefit is central to the debate. This concept is related to the measurement of the success of each partner and to the idea that the association increases this success. The success of an individual can be measured by the number of offspring that reach reproductive age. This is called fitness or reproductive success. It is often very difficult to measure. The benefit of strictly obligate symbiosis, for example, is obvious, because survival falls to zero when there is no association. But at this point, the entity formed by the partners is so interdependent that discussing the benefit to one of them in particular does not make much sense, since there is no point of comparison. Practical challenges are encountered even with facultative symbiosis. For bacterial symbionts, the difficulty of identifying, quantifying and monitoring the growth of bacteria in their natural environment renders illusory any measurement of the fitness of free-living forms, and therefore any comparison with that of symbiotic forms. Finally, the concept of benefit is offset by the cost of symbiosis: resources transmitted to the partner rather than used for oneself, for example. For similar reasons, it is extremely difficult and often even impossible to measure this cost. Besides, it should be noted that the effects of a single interaction on the partners may be beneficial or negative, depending on environmental conditions or at different times of their life cycle. These difficulties mean that symbiosis cannot be defined simply as a mutualist relationship. Indeed, in many cases, we cannot even answer the simple question: “Is this association a symbiosis?”. Anton de Bary’s original definition initially appears more usable, because it is less restrictive. However, attempting to include every type of inter-species relationship in the definition results in every association between organisms being considered symbiosis, even those that are temporary and short-lived, meaning that we lose sight of what distinguishes this particular type of association.

1.2.2. *Recent definitions: a fresh perspective on symbiosis*

In order to define symbiosis, a threshold must be determined along the continuum of varyingly close and interdependent relationships, beyond which the relationship will be defined as symbiotic. This threshold can only be artificial and it therefore makes sense to identify one or more criteria to categorize associations and define symbiosis as an object of study. Many paths have been explored; for example, many parasitologists, following on from Claude Combes in the 1980s and 1990s, have favored the concept of durable or intimate interaction. Breaking away from the concept of benefit, it characterizes the interactions between the genomes, which are established over time, focusing instead on the continuity of the link uniting the partners over the generations and its influence on their evolution [COM 95]. This concept, which originated in parasitology studies, is sometimes extended to include all symbioses, particularly as for some authors, including Combes, what we call mutualist symbiosis is merely an inverted form of parasitism, in which the host acts as a parasite to its symbionts. David Smith [SMI 87] focused on balanced interdependence between partners, which also sidelines the concept of benefit. Among the experts who have attempted to propose alternatives, Angela Douglas suggested in 1994 that the common denominator of symbioses is not the concept of reciprocal benefit, but the acquisition of a novel metabolic capability by one partner, thanks to another partner [DOU 94]. Douglas Zook, former president of the International Symbiosis Society, proposed a definition of symbiosis as the acquisition of one organism by another unlike organism and, through long-term integration, the emergence of new structures and metabolisms [ZOO 15]. These definitions imply the reproducibility of the interaction in each generation and therefore the coevolution of communication and recognition mechanisms between partners, as well as a selective advantage that ensures that the association will endure. In her lectures, Lynn Margulis used two equations to define symbiosis: "1+1=1" and "1+1>2". The first expresses the idea that two partners become one unique entity subject to natural selection (Figure 1.2). To denote this entity, formed of the host organism's cells, organelles (mitochondria, chloroplasts) and various microbial symbionts, Margulis and David Mindell suggested the term "holobiont" (Figure 1.3) [MIN 92]. This can be linked to the concept of the symbiocosm developed by Paul Nardon in France. The second equation indicates that the capabilities of the holobiont as a whole are superior to the sum of those of its individual partners. This provides a "mathematical" translation of definitions by Angela Douglas and Douglas Zook, highlighting the emergence of new capabilities.

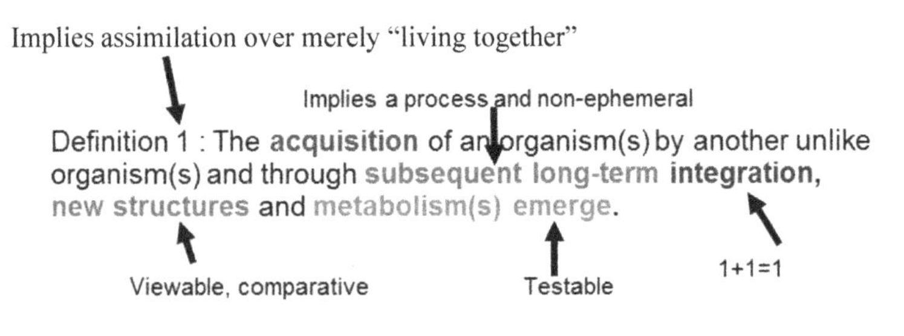

Definition 2 : 1 + 1 = 1 AND 1 + 1 > 2

Figure 1.2. *Two recent definitions of symbiosis, by Douglas Zook (definition 1 [ZOO 15]) and Lynn Margulis (definition 2 [MAR 71])*

This redefinition gives us a fresh perspective on symbiosis. These reflections are drawn primarily from the work of researchers interested in microbial symbioses, which has consequences for the application of the concept of symbiosis to interactions between other types of organisms. Indeed, many of these authors consider most classic examples of symbioses between macroorganisms to lie outside the perimeter of symbiosis today, since they are not examples of the acquisition of one organism by another or of the emergence of new structures. However, cases such as the association between ants and acacias are good examples of the latter. The tree produces sheltered structures, called domatia, and detachable food nutritional parcels, called Beltian bodies, for the ants. Once again, we are faced with the difficulty of defining the threshold at which an interaction is considered a symbiosis, in which there is a degree of subjectivity.

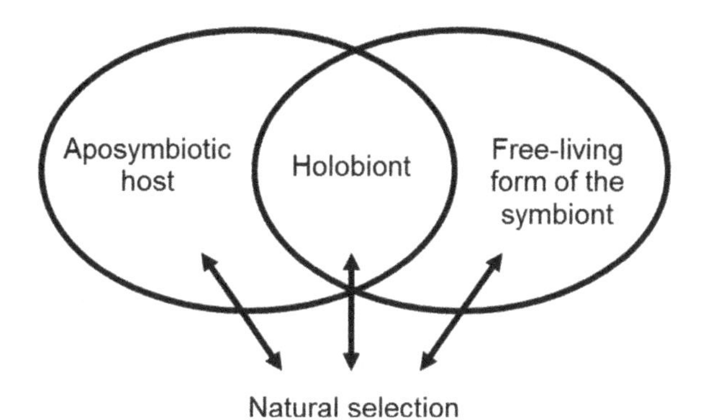

Figure 1.3. *A holobiont is a composite entity formed by a host and its symbionts. It is subject to natural selection. If they exist in nature, the aposymbiotic host (lacking symbionts) and the free-living form of each symbiont are also subject to natural selection*

The notions of cost and benefit have shaped symbiosis research for a long time and continue to do so, the main objective of many studies being to observe or measure them. But the ongoing paradigm shift is freeing researchers from this straightjacket and opening the door to a less channeled and more diverse type of research. The existence of a mutualist interaction is neither a necessary nor a sufficient condition for the existence of symbiosis. Parasitism, commensalism and mutualism must not be confused with symbiosis. They are useful for describing interactions between species within ecosystems, as Van Beneden initially proposed, but do not imply that symbiosis, as defined above, exists between them.

This significant redefinition work is gradually changing the status of symbiosis from a simple curiosity of life sciences to a genuine structured research area, as the many sessions and conferences devoted to it in recent years and the existence of a learned society (International Symbiosis Society, iss-symbiosis.org) show.

1.3. Studying symbiosis: questions and tools from past to present

Symbioses have always been discovered through observation. Over time, researchers have benefited from technical developments within the field of Life Sciences. The study of an association necessarily involves a series of questions. First of all, the partners, their characteristics and their location with respect to one another must be identified. During the 19th Century, the development and improvement of optical microscopy made it possible to identify small organisms inside larger organisms and produced the first convincing evidence of symbiotic associations.

According to convention, the larger organism was called the host and the others symbionts. Endosymbiosis, in which there is interpenetration between the partners, one inside the other, and ectosymbiosis, in which the symbiont is located on the outer surface of its host, were then distinguished from one another. The development of electron microscopy from the 1930s onwards made it possible to supplement these descriptions with ultrastructural observations at the subcellular level in subsequent decades. It was then possible to determine whether symbionts were inside the cells of their host (intracellular) or outside them (extracellular), and to locate attachment structures and specific cellular compartments (Figure 1.4).

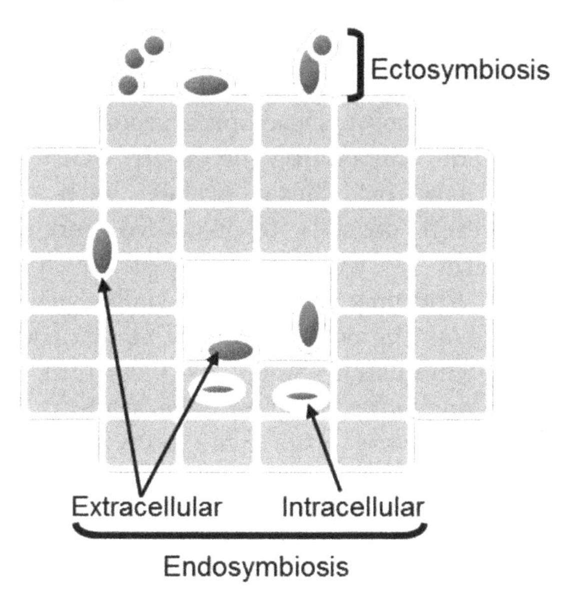

Figure 1.4. *Location of symbionts (dark gray) in relation to their host (light gray). In ectosymbiosis, symbionts occupy the external surfaces of the organism, typically its tegument, like skin bacteria. In endosymbiosis, symbionts are inside the host. They can live outside its cells (extracellular endosymbiosis) like endophytic fungi sandwiched between plant cells (see section 3.4.1), or inside the cytoplasm of the host cells (intracellular endosymbiosis), like zooxanthellae in the endodermal tissue of corals*

However, we often have to face the fact that in the world of microorganisms, seeing is not the same as identifying. Many microbes resemble others and microscopes are not sufficient, even at very high magnification. The determination of the sequence of fragments of RNA, then DNA, achieved for the first time in 1965 and 1968 respectively, became available to a large community of biologists from the end of the 1970s onwards. Methods based on sequences of marker genes, real molecular ID cards, have made it possible to identify, classify and compare bacteria in all environments. For example, they enabled Carl Woese (1928–2012) to "discover" the Archaea in 1977 by applying phylogenetic taxonomy methods to prokaryotic microorganisms [WOE 77]. PCR (polymerase chain reaction) was invented in 1985, making it possible to amplify and thus sequence fragments of DNA more easily. The development of specific labeling methods in the 1990s was another turning point. Developing probes to target DNA or RNA fragments depending on their exact nucleotide sequence and labeling these probes with fluorescent compounds made it possible to visualize and distinguish symbionts directly in the tissues of their hosts under the microscope. Proteins were made visible using labeled antibodies. This was the development of *in situ* hybridization and immunocytochemistry. Since 2005, several new high-throughput sequencing techniques have made it possible to sequence DNA on a very large scale. In some cases, it has become possible to sequence whole genomes and even all the genomes of the organisms present in a sample (metagenome). These methods are constantly improving and sequencing costs decreasing, so that we can now inventory the diversity of the gene pool of the hosts and symbionts in more and more detail. We are even beginning to be able to sequence the whole genome from a single cell of an uncultivated microorganism. At the same time, the development of high-throughput image acquisition systems is enabling us to work on whole individuals, whose every cell has been examined under the microscope.

Characterizing an association is, however, only the first step. Its day-to-day functioning must be studied, from the beginning to the end of its life cycle, to help us understand the role and contribution of each partner. This involves determining whether the association is obligate for one or the other partner, how it came about, what is being exchanged (molecules or services), and what the new properties of the system are. Experimentation therefore remains a cornerstone for the study of symbiosis. Practical difficulties often arise here. Although it is sometimes possible to physically separate partners and maintain them independently from one another in order to study their biology, the association often cannot be separated without damage. This may be because it is truly obligate, neither partner being able to survive separation, or because we do not know how to artificially create the conditions that would enable one partner to grow without the other in the laboratory. The latter is often the case for bacteria, since many of them cannot be cultivated – i.e. we have not yet managed to create an artificial culture medium that suits them. Testing hypotheses may also require one of the partners (often the bacteria) to be genetically manipulated, in order to inactivate a gene or pathway and measure its effect on the system. These restrictions mean that many fundamental questions can be addressed using only a very limited number of model organisms, those that are easiest to manipulate. The study of other associations is limited by what it is possible to do. However, technique comes to our aid for what follows. High-throughput sequencing makes it possible to deduce the capacities, functions and metabolism type of each partner, and large-scale analysis of the expression of messenger RNA (transcriptomics) or proteins (proteomics) makes it possible to compare how symbiotic systems react when exposed to the different experimental conditions imposed on them. We can observe their responses to environmental variations or identify pathways that may be involved in the communication between partners. The transfer and exchange of metabolites between partners can be traced and even quantified. Substrates with certain atoms replaced with their radioactive or heavy isotope are used to estimate the flux of matter between partners. After incubation, traces of these isotopes are sought in the various compartments of the system through mass spectrometry, or directly on histological cross-sections through micro-autoradiorgaphy or nano-SIMS, two techniques that make it possible to visually locate accumulations of unusual isotopes. We can therefore find out what these atoms have become, for example if they have first been assimilated by symbionts before being converted into constituent molecules of the host. We are starting to see the development of metabolomic approaches, which characterize all the metabolites present in a sample, i.e. the small intermediate molecules of the overall metabolism of the holobiont.

We can then measure the existing integration between the partners, ranging from a facultative and temporary relationship to a "lifelong" relationship in which the identity of each disappears into the association. There, the sequencing data are still a material of choice. They enable researchers to document the loss, gain or exchange of genes, and to estimate the speed at which genes evolve based on their mutation rates. They also allow researchers to reconstruct the partners' evolutionary relationships through comparative analysis of their genes and genomes. Evolutionary biology teaches us that encouraging the reproduction of an organism to which we are genetically related, even to our own detriment, can sometimes be an appropriate strategy. This is the theory of kin selection, popularized in 1964 by William Hamilton (1936–2000) [HAM 64]. But what if we encourage the reproduction of a symbiont with which we are not related? At any moment, one partner can attempt to divert the association to its exclusive and immediate benefit by reducing its contribution or increasing its consumption of resources. The history of a symbiosis is therefore not a long, tranquil meander, and its upkeep is not simple. In fact, symbiosis involves at least two fundamentally contradictory aspects: both cooperation and competition between partners. This is what makes it a fascinating subject of study for evolutionists.

In recent years, unprecedented amounts of data have been collected. Although this deepens our understanding of various aspects of biology, its volume requires colossal databases and computer processing capabilities. This development has led an increasing number of biologists to specialize in bioinformatics, to analyze this data, and biostatistics, to extract significant trends from it on previously unthinkable scales. This is the age of "Big Data", which affects life sciences as well as many other aspects of our societies. Progress in the study of symbioses therefore owes just as much to methodological advances as it does to new ideas. The discipline nonetheless continues to be based on three cornerstones of the biologist's work: observation, suggestion of relevant hypotheses and mastery of the tools required to test them.

2

Symbiosis and Nutrition

When symbiosis is mentioned in biology lessons, it is often through examples such as the digestion of plant matter in ruminants, the assimilation of carbon by symbiotic algae in coral or the fixation of nitrogen by the nodules of leguminous plants. Nutrition is therefore the first role of symbiosis that we think of and, although it is not the only one, it is nonetheless very important. We shall see through well-documented examples that symbiosis plays a key role in nutrition for many organisms, whether by introducing compounds such as carbon, nitrogen and minerals into the food chain, or by transforming barely digestible materials into directly usable nutrients. By using these refractory materials and recycling various metabolic waste products, symbiosis helps organisms both to save energy and to adapt to ecological niches that it alone enables them to exploit.

2.1. Becoming autotrophic

Any living organism is composed of molecules built from chains of carbon atoms. These atoms make up at least 50% of its "dry" weight, i.e. its weight once water has been removed. The carbon may come from the consumption of other organisms, whose molecules are transformed to produce the necessary constituents. This is heterotrophy, which occurs in most animals and fungi and even in many single-celled microorganisms. But the carbon can also come directly from carbon dioxide present in the atmosphere and in the oceans. Indeed, some organisms can synthesize the carbon skeleton of all their molecules using this source alone. This is autotrophy, the process behind primary production. The most obvious example is green plants. In reality, however, few organisms in nature are completely autonomously autotrophic. These are unicellular prokaryotes (lacking a nucleus), belonging to the Archaea and bacteria groups. The conversion of carbon dioxide into complex molecules consumes a very large amount of energy, which must be procured, and requires enzymes that have evolved in only a few branches of the tree of life within these groups. In eukaryotic organisms (whose cells have a nucleus), including green plants, animals and some algae and fungi, autotrophy – when it exists – always originates in a symbiosis of varying age and integration.

2.1.1. *Air and light: photosynthesis in plants and algae*

Returning to the subject of plants, photosynthesis is performed in small green compartments called chloroplasts, found, for example, inside leaf cells. Chloroplasts contain stacked internal structures called thylakoids. The membrane of thylakoids harbors photosystems I and II. Each photosystem contains a photon-collecting antenna complex that contains pigments (chlorophyll, carotenoids) associated with proteins. These absorb energy from photons and transmit it to a reaction center, which contains *a* chlorophylls and a primary acceptor (pheophytin in photosystem II). In the presence of light, an electron is transmitted from an *a* chlorophyll to the primary acceptor. In photosystem II, this electron then passes into a transport chain that produces ATP. To replace the electron that has been removed from the chlorophyll, an enzyme cleaves a water molecule to remove an electron from it, causing oxygen production. Within photosystem I, the same mechanism of transferring an electron from an *a* chlorophyll to a primary acceptor occurs in the presence of light. This time the electron is replaced by the one that was circulating in the transport chain from photosystem II. The primary acceptor yields its electron to a second transport chain, which regenerates the reducing power in the form of NADPH. Thus, photosystems use solar energy to generate chemical energy and reducing power [TAI 02], which will ultimately be used to build a new bond between two carbon atoms, one of which will come from carbon dioxide. Solar radiation therefore provides the energy needed for autotrophy. The most notable secondary effect of this lysis of the water molecule is the production of oxygen, which is indispensable for all animal life. Carbon is then fixed through the Calvin cycle. The key enzyme in this cycle, which captures the carbon dioxide, is RubisCO (ribulose-1-5-bisphosphate carboxylase/oxygenase). This is responsible for more than 99% of the primary production on our planet and is considered the most abundant protein in the biosphere [RAV 13].

As early as 1883, Andreas Schimper had observed that under the microscope, chloroplasts resembled cyanobacteria, a group of photosynthetic bacteria. Cyanobacteria are autotrophs, green or blue, and display the same type of stacked membranes as chloroplasts. Many species proliferate in most of the sunlit areas of our planet, including hot springs, rocks and the dry Antarctic valleys. Konstantin Mereschkowski, father of the symbiogenesis theory mentioned in the previous chapter, suggested that chloroplasts might be descended from cyanobacteria. He was proved right when DNA was discovered in chloroplasts at the end of the 1950s and when DNA sequences were confirmed to resemble those of cyanobacteria in the late 1970s [MCF 01]. An association with a specific bacterium from the cyanobacteria group is therefore responsible for chloroplasts and the ability of green plants to perform primary production through photosynthesis. It is estimated that the bacterium was internalized by the ancestor of a plant cell around one billion years ago (Figure 2.1). But the integration did not stop there. The genomes of sequenced chloroplasts contain no more than one or two hundred genes, in contrast to many thousands in cyanobacteria. This suggests that a large number of the genes that were originally present have been lost, specifically those that coded for proteins that are no use in their new living environment. Chloroplasts have also lost all ability to live outside their host cell. Another significant phenomenon has left its mark on the evolution of the chloroplast genome. Some genes have been transferred to the genome of the host cell, located in its nucleus. For example, one of them codes for the small subunit of RubisCO. To ensure that the RubisCO functions correctly, this small subunit produced in the host cell is sent into the chloroplast through the addition of a transit peptide, a sort of postal address that enables the cell to deliver it to the right place. This contributes to making each partner indispensable to the other and stabilizing the association.

Chloroplasts are present in many other groups beyond green plants. Both single- and multi-celled eukaryotic green and red algae work on the same principle, since they are all part of the Archaeplastida. The common ancestor of this lineage already had chloroplasts. During evolution, a unique primary endosymbiotic event, i.e. the uptake and retention of a cyanobacterium by a non-photosynthetic eukaryote host, produced all of these lineages [LEL 12]. But other eukaryotes have acquired photosynthesis in a slightly different way. Instead of acquiring a photoautotrophic bacterium, they have integrated a eukaryotic cell that already had an autotrophic symbiont into their cells. In this case, therefore, a chloroplast is present in a primary host (from primary endosymbiosis), which is itself integrated within a secondary host (secondary endosymbiosis, Figure 2.1). This kind of event has occurred many times during evolution in various groups of single-celled organisms, such as euglena, dinoflagellates and diatoms. It is recognized by the traces it leaves behind: as well as the genome and membranes of chloroplasts, membranes from the primary host are found, and sometimes a remnant of its genome in the form of a small nucleus called a "nucleomorph", which may contain some genes involved in the maintenance of the chloroplast. There are even tertiary symbioses, bringing to mind Russian dolls.

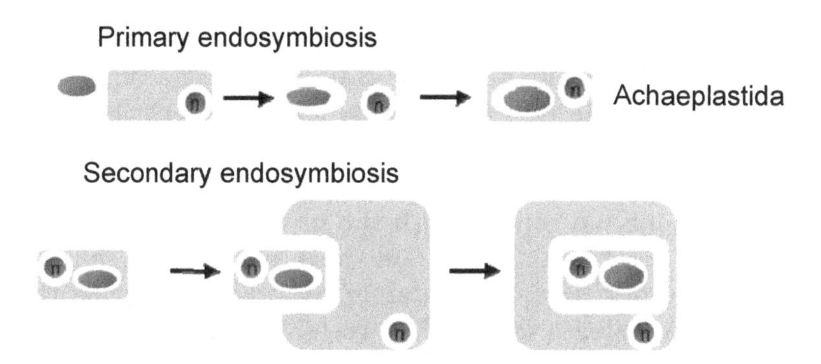

Figure 2.1. *The primary endosymbiosis of a cyanobacterium in a eukaryotic cell (gray, with a nucleus, "n") is probably a unique event that occurred within an ancestor of Archaeplastida. It is accompanied by the reduction of the genome of the symbiont through the loss of genes or their transfer to the nucleus. There are many documented occurrences of secondary endosymbiosis, in which an entire unicellular eukaryote enters into symbiosis with another host eukaryote cell. This phenomenon is often accompanied by the decay of the nucleus of the eukaryotic symbiont; the surviving remnant is then called a nucleomorph*

2.1.2. *Lichens*

Lichens, which are found in abundance on roofs, trees and seaside rocks, are a perfect example of symbiosis. It is, incidentally, not insignificant that many of the major scientists who established the study of symbiosis, including Anton de Bary, Frank and Mereschkowski, did so by working on these organisms. A lichen is made up of the combination of a fungus, the mycobiont, and a photosynthetic organism whose cells are interwoven, the phycobiont. The latter is a eukaryotic green alga (e.g. *Trebouxia*) or a cyanobacterium (e.g. *Nostoc*), and the two types can sometimes coexist. The partners are diverse on both sides, because around 20% of fungi – more than 13,500 species, notably ascomycetes – and around a hundred photosynthetic species can form lichens. The term itself simply describes the association of these organisms [LUT 09]. There are no autotrophic fungi, but the lichen becomes one through association with the photosynthetic organism, which transfers to it a significant portion of the molecules that it produces. The contribution of the fungus is to supply a stable habitat, in the form of a flattened thallus (crustose, foliose, squamulose or fruticose) that forms only on contact with the phycobiont (Figure 2.2). The thallus retains water, but also assimilates mineral elements, either from rainwater or by producing acids that attack the substrate on which it rests. The phycobiont lives inside the fungus mycelium that makes up the thallus, between the filaments (or hyphae) that surround it to increase the surface area for exchange between partners (Figure 2.3). It provides carbon molecules derived from photosynthesis, as well as vitamins. The whole forms a holobiont that is highly resistant to aridity, radiation exposure and extreme temperatures. Lichens have even survived an 18-month trip into space! Due to their resistance and very slow growth, they have an extremely long lifespan. These unusual capacities make them pioneers in difficult and extreme environments, able to colonize all terrestrial ecosystems from sea level to the highest peaks, and from humid tropical forests to hot and cold deserts. The ability to form lichens remains mysterious, but we know that it has appeared and disappeared numerous times during the evolution of fungi, mainly in ascomycetes [LUT 01].

Figure 2.2. *Lichen thalli of the Ramalina (left) and Xanthoria (right) genera. The cupules are apothecia, structures involved in sexual reproduction, on which asci develop. For a color version of this figure, see www.iste.co.uk/duperron/symbioses.zip*

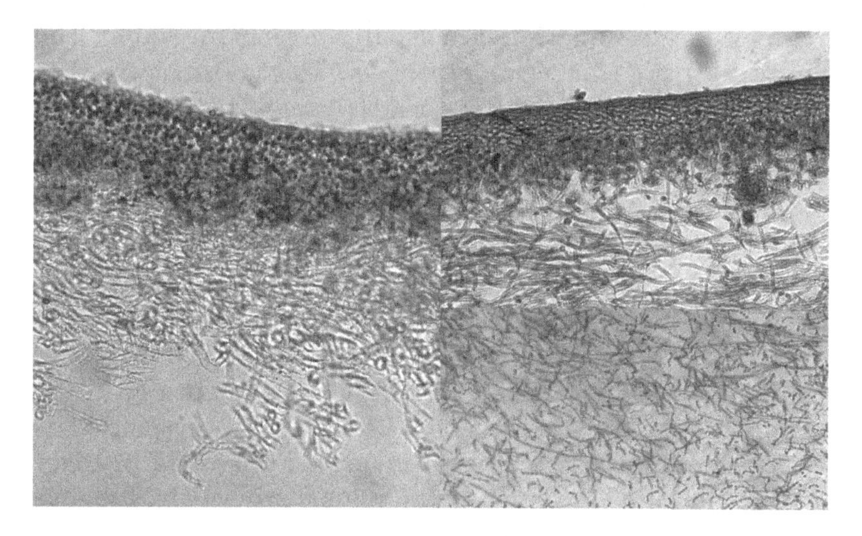

Figure 2.3. *Cross-section of lichen thalli. The general organization of the thallus can be distinguished. The upper side is flattened (upper cortex). The phycobiont cells, green on the left and brown on the right, are particularly abundant just below this area, interwoven with the filaments formed by the mycobiont (hyphae). Below, the filaments dominate (medulla), and a lower cortex is anchored to the substrate. In some lichens, the phycobiont is a cyanobacterium, which may be filamentous, such as Nostoc (bottom right). For a color version of this figure, see www.iste.co.uk/duperron/symbioses.zip*

2.1.3. *Photosynthetic animals*

Unlike plants and algae, animals are always strictly heterotrophic. However, if we consider the example of reef-building corals, the majority of their carbon comes from photosynthesis. Indeed, the endoderm cells of reef-building corals contain single-cell algae that fix carbon. Also known as zooxanthellae, they are dinoflagellates and often belong to the *Symbiodinium* genus (Figure 2.4). Symbionts are found in the transparent and well-exposed parts of the organism, which enable optimal light capture to the extent that animal cells must protect themselves from sunburn by producing molecules that filter UV rays! Host corals use various mechanisms, including proton pumps, membrane exchangers and the enzyme carbonic anhydrase, to concentrate large quantities of inorganic carbon in their tissues, up to 30 times greater than the quantities normally present in water. The symbionts use it for photosynthesis. They also recycle the carbon dioxide produced by coral tissue respiration. Thus, the algae effectively produce their own organic matter. They release 90% of it to the host in the form of small molecules, such as sugars and even some amino acids that the animal cannot produce. The host provides the nutrients they need, such as carbon dioxide and nitrogen, which seems to regulate the density of symbionts. Large quantities of carbon are fixed in this way. This productivity is vital for the ecosystem, because around 20% of the carbon is then used to form the calcium skeleton of the reef [JOH 11]. This association is obligate for the host but less integrated than the primary and secondary endosymbioses discussed previously, because it is facultative for the algae and does not involve the exchange of genetic material or proteins. Corals are not the only animals to have established symbioses with algae to feed themselves. Many species of sponge, jellyfish, hydra, flatworm, bivalve (e.g. *Tridacna* clams), nudibranch and ascidian work in more or less the same way. It should also be noted that many unicellular eukaryotes among the foraminiferans, ciliates and radiolarians, for example, also associate with symbiotic algae and make up a significant portion of marine plankton.

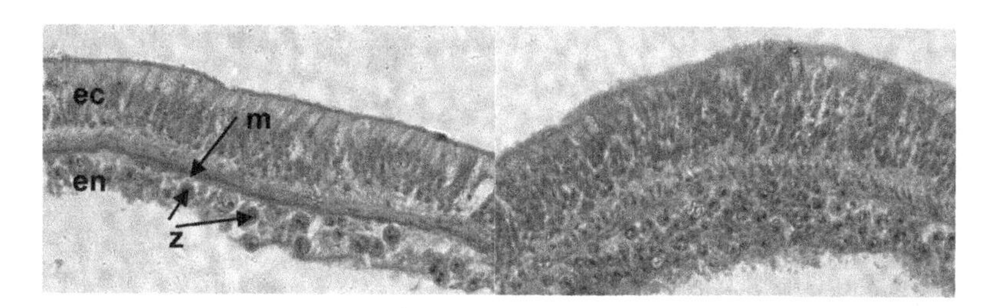

Figure 2.4. *Cross-section of a sea anemone (Anemonia sp.). The outer layer of cells, called the ectoderm (ec), lies on top of a middle layer containing very few cells, called the mesoglea (m), and an internal layer, called the endoderm (en), whose cells contain spherical zooxanthellae, here dinoflagellates of the Symbiodinium (z) genus. Although the organization remains the same, the thickness of the layers may vary, as the two images show. For a color version of this figure, see www.iste.co.uk/duperron/symbioses.zip*

But some species go even further. Like many marine invertebrates, the sea slug *Elysia chlorotica*, or eastern emerald elysia, usually feeds on algae. However, during digestion, the chloroplasts of these algae are not destroyed but captured and their function maintained for up to several months, i.e. the average natural lifespan of the sea slug. To achieve this, they are inserted into cells located in the animal's highly branched digestive tract. The sea slug can therefore live on air and light, like a plant. This type of organelle theft is called kleptoplasty (theft of plastids). The sea slug can use the chloroplasts of only two species of algae of the *Vaucheria* genus, which implies a certain degree of specificity. We do not know how the interaction between the "host" cells and the chloroplasts works during their operation period. It was once assumed that some genes had been transferred from the chloroplast genome to the sea slug genome and that they helped maintain the functioning of the chloroplasts, but this has been questioned in a recent genomic study (see section 5.6 and [BHA 13]). The fact that chloroplasts do not divide and are not transmitted to the progeny raises the question of whether or not this should be considered symbiosis [RUM 11].

2.1.4. *Chemosynthetic animals*

Light energy is the source of the majority of the primary production that we see around us. However, it is not the only possible source, nor even the simplest to use, since it relies on particularly complex machinery. One alternative is to directly use the energy stored in some reduced chemical compounds to gain the electrons needed to build new bonds during redox reactions. This chemosynthesis is less visible, but it enables some bacteria and Archaea to ensure primary production in many environments that light cannot reach, such as deep sea beds, silt and even caves (Figure 2.5). It occurs in spectacular fashion in some specific environments of the deep ocean, such as hydrothermal vents and cold seeps. There, high biomasses of large animals – giant annelid worms, bivalves, gastropods and shrimp – thrive in the darkness, around chimneys that emit burning, toxic fluid or around methane emanation sites (cold fluids). The reason for this is simple. The reduced compounds, usually present in the anoxic phase and therefore in the subsurface, mingle in the seawater, which contains oxygen that the prokaryotes use as an electron acceptor. Some prokaryotes, called chemoautotrophs, possess molecular machinery that enables them to gain large quantities of energy from this rare coexistence of good electron donors and good acceptors, which they then use to fix carbon and transform it into organic matter. The carbon assimilation pathway is usually the same as in photosynthetic organisms, i.e. the Calvin cycle based on RubisCO, but there are other pathways, notably the reverse tricarboxylic acid cycle (reverse Krebs or "rTCA cycle"). In the case of hydrothermal vents, the fluids result from the interaction between seawater and the newly formed oceanic crust, which it penetrates before heating up, due to magmatic activity at ocean ridges. This is a geological phenomenon. Fluids at cold seeps have diverse origins, generally related to accumulations of buried organic matter at the foot of continental slopes. The reduction of this organic matter through the action of the temperature or of methanogenic and sulfate-reducing prokaryotes causes it to be converted into hydrocarbons, including methane, and other reduced compounds that come from the biological activity of microbial consortia such as dihydrogen sulfide. Symbioses with chemosynthetic bacteria play a key role in all these ecosystems, because most animals found in abundance there depend on them for all or part of their nutrition [DUB 08]. In the giant tube worm *Riftia pachyptila*, which can grow to over 2 m in length, bacteria use hydrogen sulfide as an energy source and are located within the cells of the trophosome, a nourishing organ that accounts for most of the animal's volume (see Figure 5.2). The adult worm has no digestive tract and takes all its nutrients from the symbionts and their digestion on the periphery of the trophosome. In vesicomyid bivalves (e.g. *Calyptogena*) and mytilid bivalves (e.g. *Bathymodiolus*), which can grow to over 30 cm in length, most symbionts use sulfur. Some symbionts associated with a number of mytilidae use methane. Bivalve-associated symbionts are found inside the specialized cells that form the epithelium of the hypertrophied gills (Figure 2.6). Their hosts' digestive tube is generally reduced, again emphasizing the prominence of symbiotic nutrition [DUP 13]. These symbioses have appeared independently many times within various lineages of animals and bacteria, and are an important characteristic of these habitats. Species presenting similar symbioses have been found in environments closer to us, such as coastal sediment and mangroves, where the presence of reduced compounds is also linked to the decay of organic matter, but they do not play such a key role in the productivity of these ecosystems.

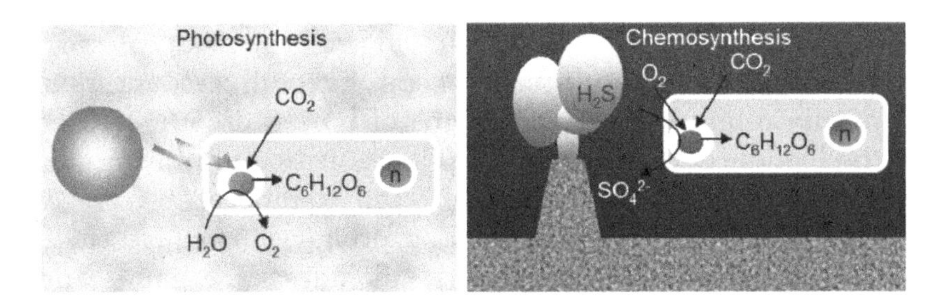

Figure 2.5. *Two types of primary production. Oxygenic photosynthesis (here performed by the chloroplast of a eukaryotic cell) uses light as an energy source, water as an electron donor and carbon dioxide as a carbon source. It produces sugars as well as oxygen. Chemosynthesis (this example shows an intracellular sulfur-oxidizing bacterium in a eukaryotic cell) uses chemical energy released by the oxidation of reduced compounds such as dihydrogen sulfide (an electron donor, here emitted by a submarine hydrothermal vent) in the presence of oxygen. This energy is used to transform carbon dioxide into sugars, and sulfate is produced*

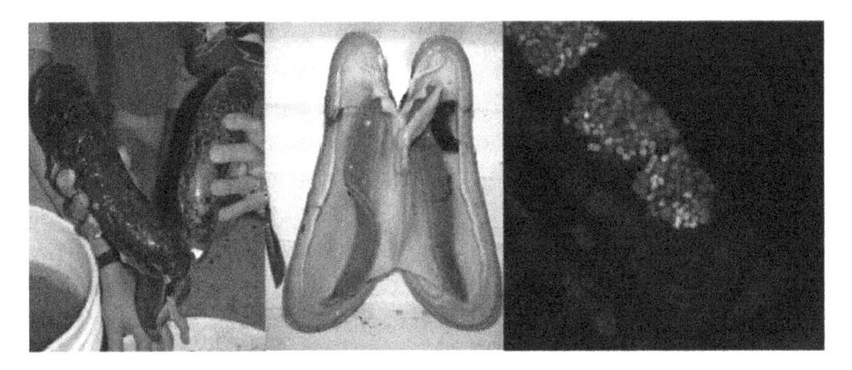

Figure 2.6. *Mussels of the Bathymodiolus genus sometimes reach considerable sizes (left). They are characterized by hypertrophied gills (brown, center), which contain abundant populations of symbiotic bacteria. The image on the right shows in-situ hybridization performed using fluorescent probes, on which we see a few cells of the gill epithelium, their nuclei blue, containing both methanotrophic bacteria (green) and sulfur-oxidizing bacteria (pink). For a color version of this figure, see www.iste.co.uk/duperron/symbioses.zip*

2.2. Assimilating nitrogen and other elements

Carbon is not the only element that is essential to life. Hydrogen, oxygen, nitrogen and many other elements (calcium, phosphorous, potassium, sulfur, sodium, metals, etc.) are required in smaller quantities. Hydrogen and oxygen are abundantly present in air and water, but the others are harder to procure. Once again, association with symbionts is vital for many organisms.

2.2.1. *Bacterial fixation of atmospheric nitrogen in plants*

Nitrogen makes up 78% of the air that surrounds us. Procuring it would pose no problem to organisms if atmospheric nitrogen, or dinitrogen, were not an inert gas, meaning that it is almost impossible to separate it into atoms that organisms can use. This is due to the triple bond that binds the two nitrogen atoms together. The decay of dead organisms provides nitrogen sources that can be assimilated in the form of ammonia, nitrates, amino acid derivatives and nucleotides, which are typically used by heterotrophic organisms, but simple recycling is not enough: the nitrogen must first enter the food chain.

To become assimilable, atmospheric nitrogen must first be transformed. Organisms fix it by means of an enzymatic complex called nitrogenase, which transforms nitrogen into assimilable ammonia. This reaction requires a lot of energy, taking into account the stability of dinitrogen, using 16 molecules of ATP to convert a single molecule of dinitrogen, and nitrogenase tolerates only very low concentrations of oxygen, after which it becomes inactive. Only a few species of bacteria and Archaea naturally possess nitrogenase, and no eukaryotic organism, plant or animal, has it in its genetic arsenal. It is, however, well known to farmers and gardeners that soya, peas and beans make soil nitrogen rich, and many species, such as melilotus and vetch, are used as green manure. Once again, it is symbiotic association that has given some plants this ability. These plants are Fabaceae or legumes and have small bulges called root nodules on the surface of their roots, as far as possible from leaf oxygen production areas. These structures contain bacteria called Rhizobia, which actively fix nitrogen. The most studied belong to the *Rhizobium* genus (Figure 2.7). In reality, Rhizobia is a collective name for more than one hundred different species, classified in at least 12 genera of Alpha- and Betaproteobacteria [REM 16]. In normal circumstances, these bacteria live in the soil and do not express nitrogenase. When placed near their host's root, they multiply in response to its emission of compounds that encourage bacterial growth, establish themselves in the root, which develops a nodule, and multiply there (for details, see section 4.1.1). Once inside the root nodules, the symbionts benefit from a very controlled environment. For example, leghemoglobin, a specific hemoglobin produced by the plant, restricts the quantity of oxygen present, thus preventing the inactivation of the nitrogenase. The bacterium receives carbon compounds, such as malate, from the plant in order to feed itself, and in exchange provides nitrogen in the form of ammonia, which the plant can use. Many other non-leguminous plants, such as heather and alder, associate themselves with nitrogen-fixing bacteria of the *Frankia* genus and also form root nodules.

Some ferns, cycads in their coralloid roots (Figure 2.8), and even the spectacular *Gunnera*, associate with cyanobacteria, such as *Nostoc* or *Anabaena*. We have already examined the photosynthetic role of these cyanobacteria when they associate with lichens. Of course, in plants, this aspect is of comparatively minor interest, because they practice photosynthesis themselves on a large scale. What is interesting is the ability of cyanobacteria such as *Nostoc* to fix atmospheric nitrogen [RAI 00]. This is rather paradoxical, because photosynthesis releases oxygen, which is detrimental to nitrogen fixation. In fact, the problem is circumvented, because some of the cyanobacteria specialize in fixing nitrogen by abandoning photosynthesis entirely and receiving their carbon from neighboring cells or from the host. Some cyanobacteria fix nitrogen in lichens as well. Later on, we shall see that symbiosis also enables some animals to assimilate atmospheric nitrogen during digestion.

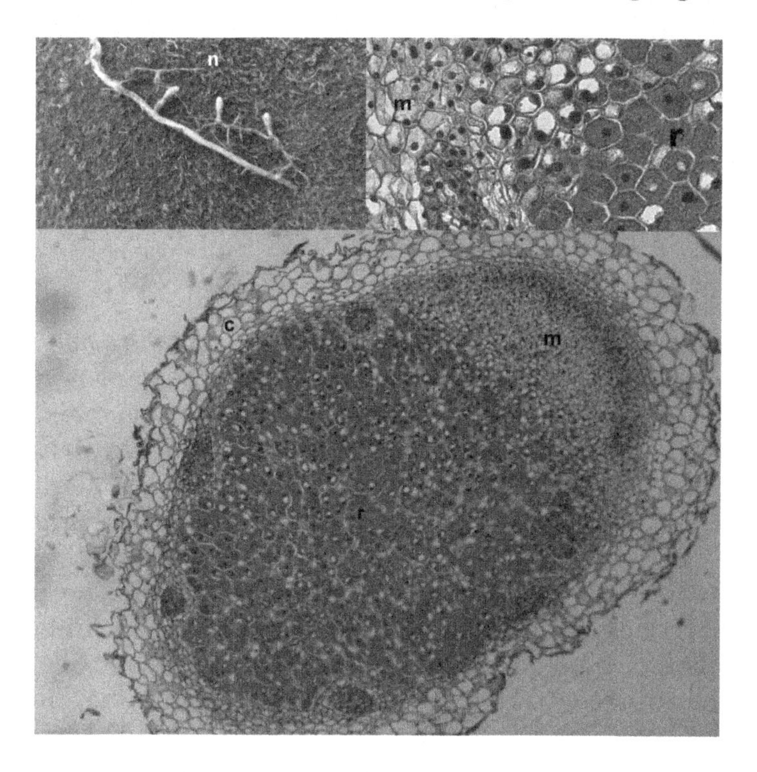

Figure 2.7. *Top left, root nodules (n) on an alfalfa root (Medicago sp.). The pinkish color is due to the presence of leghemoglobin. Below, cross-section of a nodule showing the root cortex on the edge (c), the area occupied by the nitrogen-fixing Rhizobia, in the middle (r), and the apical meristem, the growth area of the nodule, composed of small cells without bacteria (m). Top right: detail. For a color version of this figure, see www.iste.co.uk/duperron/symbioses.zip*

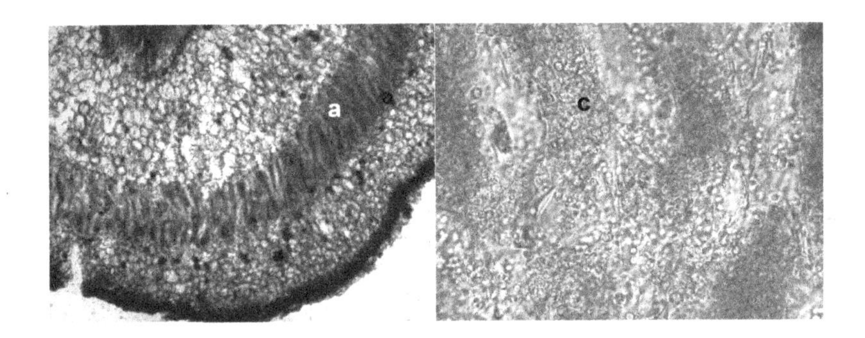

Figure 2.8. *Cross-section of a coralloid root of Cycas revoluta. Inside the root is a visible ring, left (a); the detail on the right shows that it contains many cyanobacteria (c). The role played by cyanobacteria in fixing atmospheric nitrogen is similar to that of Rhizobia in the root nodules of other plants*

2.2.2. *Access to soil elements by mycorrhizal fungi*

Many other elements are necessary for life, although they make up less than 1% of the dry weight of an organism. They include potassium, calcium, magnesium, phosphorous and many others. In terrestrial green plants, they are absorbed from the soil, mainly through the roots. In the majority of these plants, however, roots act in close collaboration with fungi. These fungi appear as filaments (or hyphae), which cover the roots and spread out into the ground. The plant provides them with compounds, such as simple sugars and some vitamins, exuded through its roots. The metabolic cost is high: transfer rates of around 15–30% of the carbon derived from photosynthesis have been observed [FIN 08]. This enrichment of the root environment in easily exploitable compounds encourages the growth of the fungus, which is heterotrophic. It produces hyphae that form a network that physically extends the roots. This establishes an exchange interface with the soil that is 100–1,000 times larger than that of the roots alone. Through this surface area, the fungi absorb and channel a large number of elements from the soil.

These include phosphorous, copper, iron, manganese and water. A portion of these elements goes back to the plant. This association is most beneficial in poor soils and most efficient when it comes to taking in elements that are particularly immobile, such as phosphorous, which the fungi mobilize by excreting phosphatases. In terms of nitrogen, the fungi can remobilize what is contained in the decomposing organic matter, but the presence of mycorrhizae also facilitates the nodulation of roots by nitrogen-fixing bacteria.

Like lichens, mycorrhizae are not organisms but the product of an association between the roots of a plant and the mycelium of one or several fungi. The term, coined by Albert Frank in 1885, is composed of the Greek words *myco* (fungus) and *rhiza* (root). The hyphae are thin filaments capable of covering the root like down. For example, the truffle is the reproductive part, or carpophore, of a mycorrhizal fungus whose mycelium is associated with trees. In the most common type of mycorrhiza, found in around 80% of plants, the fungal hyphae push extensions through the plant cell walls between them and the plasma membrane. These structures are called arbuscules because of their tree-like shape. This is arbuscular endomycorrhiza (Figure 2.9). The exchange surface between the plant cells and the fungus is increased and exchanges facilitated. There are other forms of endomycorrhiza in the heather family (Ericaceae) and in orchids (Orchidaceae). Some plants, including Ericaceae, such as *Monotropa*, and Orchidaceae, such as *Neottia nidus-avis*, have even lost their chlorophyll and become entirely dependent on the fungus, even for their carbon supply. In this case, the fungus provides the carbon by connecting itself to the mycorrhizal network of other plants. These plants are called myco-heterotrophs. In orchids, mycorrhizae also play an important role in the supply of energy during the germination of seeds, which are minuscule and therefore have almost no supplies.

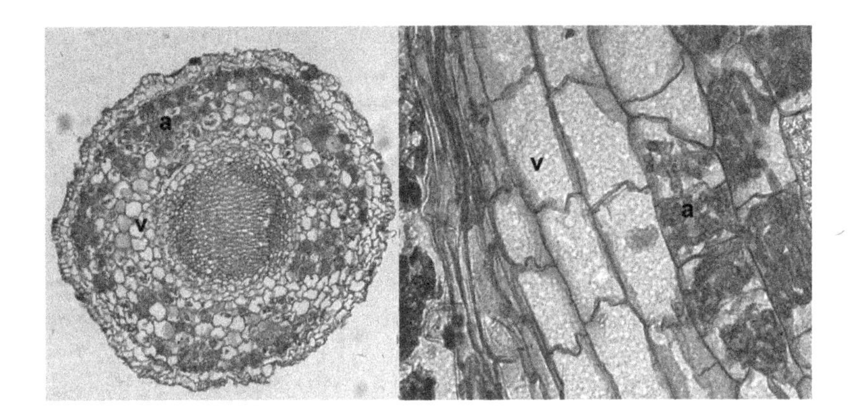

Figure 2.9. *Cross-sections of a mycorrhizal ivy root. Plant cells and their large, transparent vacuole can be distinguished (v). Hyphae penetrate the root tissue, passing through the cell walls and between membranes. They branch off at the end, forming arbuscules (a) in some of the plant cells, increasing the exchange surface between the partners. At the other end, the hyphae stretch out in the soil and form a network that cannot be seen here. The fungus belongs to the glomeromycetes group and this type of mycorrhiza is called an arbuscular endomycorrhiza. For a color version of this figure, see www.iste.co.uk/duperron/symbioses.zip*

In some trees, such as beeches, oaks, birches and conifers, the fungal hyphae penetrate the root in a limited way by squeezing between the cells to form a Hartig net, but do not penetrate the cellulose wall (Figure 2.10). This is ectomycorrhiza, which occurs in around 5–10% of plant species but is ecologically very successful, since these plants cover large areas of boreal forest.

Mycorrhizae can be considered obligate symbioses, since it is difficult for the partners to survive without one another. The association is, however, not very specific: a given plant can often establish a relationship with many dozen species of fungi at the same time. A few hundred species of fungi are known to be capable of forming endomycorrhizae, although the difficulty of distinguishing between species means that this is probably an underestimation.

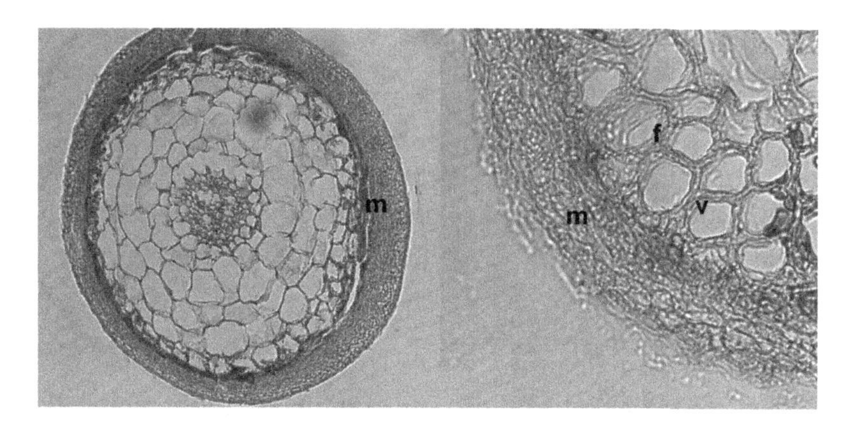

Figure 2.10. *Cross-section of a mycorrhizal oak root. An outer sheath (m) formed of hyphae (fungus filaments) surrounding the root can be seen on the left. On the right, the detailed view shows the filaments (f) interwoven between the plant cells (v) without penetrating them, forming a Hartig net. This type of mycorrhiza is called ectomycorrhiza*

2.3. Digesting food

Unlike plants, animals and heterotrophic organisms generally consume complex molecules and need only to transform them or collect the building blocks they need to live from them (sugars, amino acids, etc.). Nevertheless, the digestion of the alimentary bolus in animals involves the participation of communities of microorganisms. In vertebrates, these communities are so complex that they become difficult to study. Some invertebrates, however, have simpler systems that are easier to study from a practical point of view because they have short life cycles and fairly unrestrictive breeding conditions. Many studies have therefore focused on bacteria in the digestive tracts of invertebrates. Among them, the most frequently studied are animals in which a limited diversity of microbial partners leads to their adaptation to a remarkable or astonishing alimentary regime. In principle, an animal consumes proteins, lipids and carbohydrates. However, in reality, many animals have an unbalanced diet. Symbiosis can help to reestablish this balance and to exploit food that is poor or lacking in diversity. Thus, termites eat wood, aphids absorb plant sap, and leeches feed on the blood of vertebrates. Recently, tubeworms have even been discovered feeding on collagen from whale bones on the ocean floor [GOF 05]! Whether the primary matter is resistant to digestion or low in nutrients, or even both, these diets are, in theory, very negative.

2.3.1. *Wood-eating termites*

Wood is made of cellulose and lignin, complex molecules that are very difficult to break down, and also contains very little nitrogen. The digestive tract of termites has adapted so that they can feed themselves. "Lower" termites shred the wood into small pieces and produce a few enzymes to break down the cellulose, but the quantities are far too small to digest all of it [BRU 14]. A large specialized paunch has developed in the hindgut, a region of the digestive tract that is particularly low in oxygen. It contains up to several million unicellular flagellate eukaryotes [OHK 08]. Their flagella enable them to move and they have a repertoire of enzymes (cellulases and hemicellulases) that work in low-oxygen conditions, meaning that they can digest large quantities of lignocellulose. Flagellates release short-chain fatty acids to the host (Figure 2.11). But this is not all. There are also many hundreds of species of bacteria and Archaea living in the digestive tract of termites, especially on the surface and within flagellates, forming "nested" symbioses. Although all their functions are not yet known, they include species capable of fixing atmospheric nitrogen like the symbionts of leguminous plants, breaking down cellulose (perhaps both), and transforming sugars into compounds such as amino acids and cofactors, which are essential for flagellates, or into acetate, which termites use as a carbon and energy source [HON 08]. "Higher" termites have no flagellates in their gut and the prokaryotes break down all the cellulose themselves [WAR 07]. These complex associations contribute to supplementing and balancing out the diet of xylophagous termites. Another group of termites, called "fungus-growing", is unable to digest wood in its digestive tract, so outsources the work. These insects cultivate a fungus called *Termitomyces* inside their termite mounds, on piles of chewed-up but undigested vegetable matter. *Termitomyces* breaks down the lignin and cellulose, whereupon the piles are consumed and assimilated by the termites. Although it is unusual, xylophagy based on symbiosis with bacteria (similar to that practiced by "higher" termites) does exist in other species, for example in marine mollusks, of which the most common are shipworms, bivalves that cause significant damage by attacking submerged wood, such as boat hulls and bridge piers. They use their grainy shells to shred the wood and digest the shavings with the help of cellulolytic bacteria, which are also nitrogen fixers [LEC 07].

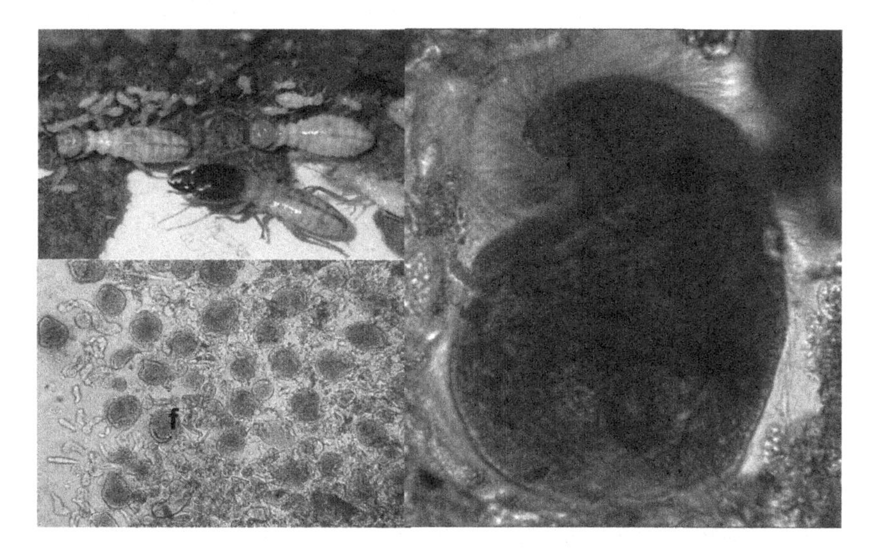

Figure 2.11. *"Lower" termites, such as Hodotermopsis sjoestedti (top left), contain dense populations of flagellates (bottom left, f) in a paunch in their digestive tract. Right: detail of a flagellate of the Trichonympha genus. Photos reproduced with permission of D. Sillam-Dusses. For a color version of this figure, see www.iste.co.uk/duperron/symbioses.zip*

2.3.2. *Sap-sucking insects*

Many insects feed on plant sap. For example, aphids use their mouth parts as a straw, sticking them into the phloem vessels of plants to suck out the elaborated sap, which contains sugars and minerals, but none or very few proteins and lipids. Although aphids do not become diabetic, this diet nonetheless causes problems. Aphids produce only 10 of the 20 main amino acids that make up proteins, and must therefore find the 10 others in their food. Like many insects, they contain a bacterial endosymbiosis, in this case involving a bacterium of the *Buchnera* genus. Since aphids are easy to procure and breed and reproduce very quickly, the aphid-*Buchnera* pairing has become one of the best-studied models of symbiotic association. These studies have led to many discoveries that have wide implications for symbiosis. The bacterium is located within specialized cells (bacteriocytes) of two symmetrical organs called bacteriomes, which are found only in the abdomen of symbiotic insects [DOU 98]. *Buchnera* receives sugars from its host, produces the 10 essential amino acids that the host cannot synthesize and transfers them to it. The symbiosis is obligate for both partners. An aphid treated with antibiotics and therefore lacking symbionts can no longer grow or reproduce, and *Buchnera* does not survive when it is removed from its host.

The study of the *Buchnera aphidicola* genome has revealed the origins of the bacterium's dependence on the aphid. The genome contains a maximum of around 640 kilobases (1 kilobase, or kb, corresponds to 1,000 base pairs), as opposed to 4,000–5,000 kb for a comparable bacterium living in the environment [SHI 00]. This genome reduction causes the disappearance of a large number of genes. Fewer than 600 remain, as opposed to 5,000 for an equivalent free-living bacterium [DAL 06]. The genes for defense, regulation in response to environmental variations, anaerobic respiration, and those coding for various extracellular structures that are present in free-living bacteria have been lost, attesting to *Buchnera*'s long isolation within highly controlled host cells. Many genes considered vital for DNA repair and recombination are also absent. Most remarkable is the loss of genes enabling the production of amino acids that are already produced by the host. Each partner seems to specialize in the production of one portion of the amino acids, and is complemented by the other. The tasks seem to be shared and the reciprocal dependence reinforces the temporal stability of the association and the control between partners. This genome reduction is also linked to the mode of bacterial transmission: as we shall see, they pass from one generation of aphids to the next through eggs during sexual reproduction, or through tissues during asexual reproduction, without ever encountering the environment.

2.3.3. *Blood-sucking insects*

The tsetse fly (*Glossina* genus), a known carrier for sleeping sickness, feeds on the blood of vertebrates, another highly unbalanced alimentary regime. Like the aphid, this fly contains one main symbiont, *Wigglesworthia*. This is an obligate symbiont found only in specialized host cells and its genome is reduced, like that of *Buchnera* (to around 698 kb). The bacterium has lost many genes, including those required for the production of many amino acids. *Wigglesworthia*'s role is to complement its host's alimentary regime by producing vitamin cofactors, mainly from group B [WER 02, SNY 13]. These vitamins, particularly B6, are vital for the production of proline. The *Glossina* fly uses proline as an energy source. Furthermore, it is viviparous. Its embryos develop inside the uterus and are fed a mixture of lipids and proteins secreted by the mother. To fuel this system, which uses a lot of resources, the female must produce a lot of proline. In fact, without *Wigglesworthia*, the females are infertile. The mother transfers the symbiont *Wigglesworthia* through the glands that feed the embryos. A second important symbiont, *Sodalis*, is present but does not appear to complement *Wigglesworthia*. On the contrary, it seems to have redundant functions, such as the production of vitamin B.

2.3.4. *The gut microbiota of vertebrates*

The digestive tracts of all animals contain many microorganisms, whose role in digestion has long been underestimated. The development of new tools, particularly for the high-throughput sequencing of metagenomes, has enabled researchers over the last few years to inventory these agents and clarify their functions. We can then only observe to what extent the alimentary regime (herbivore, carnivore or omnivore) and evolution have left traces in the composition of these microbial communities [LEY 08]. This colossal task culminated in the sequencing and study of the human microbiome, begun in 2008, which aims to characterize all the microorganisms associated with our species, including those in the digestive tract [TUR 07, CON 12]. It is estimated today that the human body contains between 3 and 10 times more bacteria than animal cells and contains 1,000 times more bacterial genes than human genes. These bacteria are small and make up “only” 1 to 3% of our weight, but represent many hundreds and perhaps thousands of different types of bacteria [MAZ 08]. Around 90% of them are located in the large intestine and are indispensable to our well-being [FOR 10]. This microbial community helps us break down foodstuffs in order to extract nutrients from them, including minerals such as iron. It produces important substances, such as some group B vitamins that our bodies cannot produce [KAU 11]. It also helps us break down refractory vegetable matter, such as cellulose, although this specific role is much more developed in herbivores. However, it also plays many other roles that we are only just beginning to shed light on in many pathologies, such as obesity and Crohn’s disease, and even in protection against pathogens, the education of our immune system and the development of our nervous system. These discoveries open fascinating new perspectives, which will be presented in more detail in Chapter 7.

The digestive tract in herbivorous vertebrates is proportionally longer than in carnivores and omnivores, and some areas are specialized to form chambers for cellulose digestion. These chambers contain specific, complex microbial communities, associating flagellate eukaryotes, fungi, Archaea and bacteria, which digest what the host cannot [DOU 94, MAC 02]. These chambers are located in different regions of the digestive tract, according to the groups. In ruminants, for example, the rumen is located upstream from the stomach (pregastric chamber). These animals' genomes contain multiple copies of the gene that codes for lysozyme. The initial function of this enzyme, which destroys the bacterial wall, is protection against pathogens and regulation of microbial communities, but in ruminants it instead helps to digest large quantities of microorganisms that develop in the rumen, in order to assimilate nutrients from them downstream from the stomach, in the intestine. Other animals have postgastric cellulolytic symbioses located distal to the enzyme-secreting zone of the gut, in the large intestine or in chambers called caeca. Location far back in the digestive tract means that not all products of digestion can be absorbed. To circumvent this limitation, some species such as rabbits consume their own feces to assimilate the products of the fermentation (coprophagy).

2.4. Recycling waste

Nitrogen is part of the composition of many biomolecules, such as amino acids, which make up proteins, and bases, which make up DNA and RNA. The general metabolism therefore produces significant quantities of nitrogenous waste in many forms, such as ammonia, urea and uric acid.

For some organisms, notably those that feed on a material that is naturally low in nitrogen (e.g. plants), the loss of such an important element can cause problems. Some examples of symbioses that recycle the nitrogenous waste of animals are well documented. In corals, nitrogenous waste from the host's digestion of animal prey is a nitrogen source for the symbionts within the host's cells, which also absorb ammonia and nitrate in the water. In animals equipped with a urine evacuation system, waste-recycling symbionts must occur in the ducts where it is stored. In insects, primary urine is produced in the Malpighian tubules, which act as kidneys and open onto the digestive tract, so that urine passes through it before being evacuated. Some bacteria present in the digestive tract of insects therefore specialize in recycling the components of this waste, primarily uric acid. This is observed in xylophagous termites, whose alimentary regime is particularly low in nitrogen. They contain many types of bacteria that can use uric acid, reducing its quantities to almost nothing. The bacteria convert the nitrogen waste into ammonia for their own sake, which also benefits the animal [THO 12]. In cockroaches, the bacterium *Blattabacterium* is located in cells that store fats and uric acid. *Blattabacterium* cannot use this acid directly and we do not yet know whether it is the animal or another as-yet-unknown digestive bacteria that transforms it into ammonia or urea. Nonetheless, *Blattabacterium* uses these two compounds, and uses the nitrogen to produce the 10 amino acids that are essential to its host [SAB 09]. Waste recycling by symbionts is thus of great importance to animals whose alimentary regime is low in nitrogen.

3

Symbiosis and Other Functions

Besides ensuring its nutrition, the life of an organism also consists of interactions with other individuals of its species (e.g. during reproduction), with different species (prey, predators, competitors, mutualists and pathogens) and, of course, with the environment. In animals, the nervous system, the sense organs and even the locomotive system are particularly important for these interactions. Just as for nutrition, the organism does not act alone and symbionts often participate actively in these functions. This chapter will show how symbiosis has caused new functions to emerge, such as controlled bioluminescence in some animals and mobility in some protists, and how it helps improve reproductive performance and protection from hostile conditions for a large number of species. Although cited less often than nutritional symbioses, these symbioses are just as important for the survival of the host.

3.1. Seeing, being seen, hiding: bioluminescence

In recent years, strange species have appeared on the counter at the fishmonger's. As resources are gradually exhausted, we fish deeper and deeper for largehead hairtail, ling and even grenadier. The sun's light does not reach these depths. However, if the head has not been cut off (commonly done to prevent the customer's appetite from being ruined by this disgusting "appendage"), we can see that these species have large eyes. Eyes to see nothing? In reality, in the deep ocean, the darkness is studded with points and flashes of light. Around 90% of the organisms found there produce light, continuously or intermittently, at varying wavelengths [HER 87]. However, it is not in a deep-sea species that this system has been most effectively analyzed today, but in a small sepiolid squid from Hawaii, a few centimeters long and easy to keep in the laboratory, called *Euprymna scolopes* (or "bobtail squid"). On its ventral side, the squid has a light organ that resembles a pair of headlamps. There are two lobes, each consisting of three crypts containing the bacterium *Vibrio fischeri* (recently renamed *Aliivibrio fischeri*), which creates light (Figure 3.1). The basic principle of bioluminescence is very simple: it requires an enzyme classified as a luciferase (light-producing enzyme) and a substrate called luciferin. When the substrate is oxidized by the enzyme, a large quantity of energy is emitted. This energy takes the form of a visible photon and therefore a flash. If a single bacterium glows, the effect is not particularly spectacular; in fact, a single bacterium will not glow.

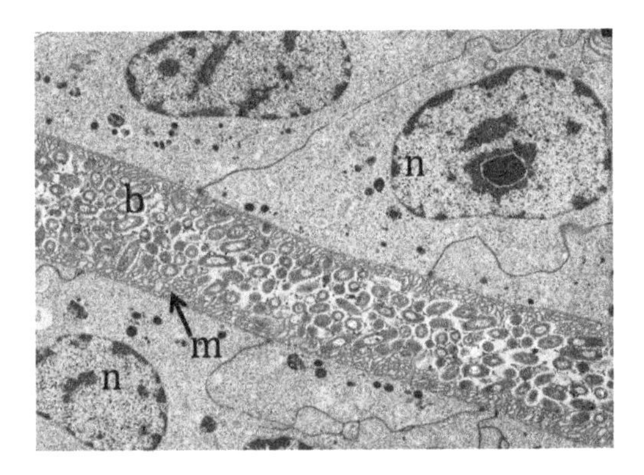

Figure 3.1. *Detail seen under transmission electron microscope of a crypt containing Vibrio fischeri (b). The crypts are bordered by epithelial cells (n: nucleus), which emit extensions called microvilli (m) towards the area containing the bacteria. Photo reproduced with permission of M. McFall-Ngai*

In *Vibrio fischeri*, light production is controlled by complex molecular mechanisms. Each bacterium in a crypt produces specific signaling substances and has receptors to these same substances on its membrane (these substances are therefore named autoinducers). Light is emitted synchronously in all the bacteria if and only if the concentration of these autoinducers exceeds a certain threshold (Figure 3.2). This threshold represents the high density of bacteria in a crypt (Figure 3.1), around 100 billion/ml, guaranteeing the visibility of the light emission. The capacity of the bacteria to communicate and to perceive one another is called quorum sensing, and we owe its discovery to studies on bioluminescent symbiosis in squid. This was one of the major biological discoveries of recent years, since what followed demonstrated the involvement of this same mechanism in many other important bacterial functions, such as mobility, sporulation, virulence and even biofilm formation [MIL 01].

In squid, the function of this bioluminescence is probably camouflage through counter-illumination (Figure 3.3). Living in shallow waters in which light penetrates easily, the animal projects a shadow that makes it easy for predators to spot, particularly those on the seabed. The light organ projects light of bacterial origin towards the seabed. Its intensity can be adjusted to match the incident light, and it is released through a lens similar to the one in the squid's eye. Light production peaks in the evening, while the animal feeds in starlit water, and 90% of symbionts are evacuated at dawn, meaning that light emission ceases just as the animal returns to hide in the sand for the day. The *Vibrio* quickly begin to multiply again and the density reaches the threshold required for light emission to begin the next evening [MCF 14a]. This symbiosis, therefore, works on a daily rhythm. *Vibrio fischeri* has no need of its host and survives well in seawater. However, in associating itself with the light organ, the bacterium gains a very favorable environment. Indeed, the walls of the light organ secrete sugars and amino acids for bacteria to feed on, enabling them to multiply more rapidly. The crypts act as bioreactors for *Vibrio*, accelerating its growth.

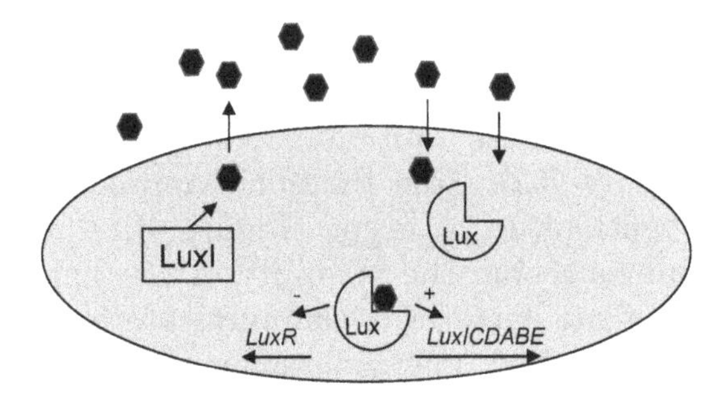

Figure 3.2. *Quorum sensing in Vibrio fischeri. The LuxI protein is responsible for the production of the autoinducer (here an N-homoserine lactone, represented by hexagons). As bacterial density increases, the concentration of autoinducers increases outside and inside the bacterium. When the critical concentration has been reached, the autoinducer attaches itself to the LuxR protein. The resulting complex binds itself to the promoter of the luxICDABEG operon and activates its transcription, suppressing that of LuxR. The operon codes for the luciferase responsible for light production (LuxAB), as well as a protein that enables the production of luciferin, a substrate of the reaction (LuxCDE). The LuxI protein also becomes increasingly abundant, increasing autoinducer production and therefore ultimately light production. At the same time, LuxR becomes less and less abundant, preventing the system from becoming overactive. Diagram adapted from [MIL 01]*

Figure 3.3. *Principle of counter-illumination. In the squid Euprymna scolopes (green and yellow), the ventral bioluminescent organ produces light, masking the shadow that the animal would normally project (right), hiding it from potential predators. For a color version of this figure, see www.iste.co.uk/duperron/symbioses.zip*

In bacteria of the *Vibrio*, *Aliivibrio* and *Photobacterium* genera, bioluminescence is related to the presence of a conserved *lux* operon that codes for the synthesis of the luciferase enzyme and its substrates (Figure 3.2). Although bioluminescence can be attributed to symbiotic bacteria in the majority of light-producing animals, some produce bioluminescence themselves, through a wide range of enzymes and substrates. Examples include glow-worms, some fish and squid, cnidarians, ctenophores and arthropods, as well as various fungi and algae. Many chemistries have been characterized and bioluminescence has appeared around 40 times in the course of animal evolution [WID 10]. Its role is not confined to camouflage. Fish such as lanternfish and anglerfish use it to lure their prey close to their mouths. In anglerfish, bacteria are contained in a contractile organ. When it contracts, bacterial density increases. Beyond the density threshold, light is produced, and when the organ relaxes, production ceases. Another important function is lighting the visual field, as in the fish *Anomalops katoptron*. The areas containing bacteria, beneath its eyes, are not unlike headlamps. Bioluminescence also helps to attract partners for reproduction and scare off predators; one deep-sea vampyromorph octopus emits a cloud of light particles to disturb its predators. In this case, the light is not symbiotic but is produced when luciferin and luciferase, which are stored in separate glands, are emitted simultaneously into the environment.

3.2. Movement: phoresy

Phoresy is a type of relationship in which one partner is transported by another. A well-known example is the transport of mites by various arthropods, particularly insects, from which they never separate themselves. There are many examples of this in a more general sense. Although the acquisition of a new capacity by one partner through an association is involved, movement is usually a secondary effect of an interaction and not the main reason for its continuation. We have perhaps reached the limit of what our initial definitions consider to be symbiosis. Another frequently studied example is perhaps more convincing, since the symbionts play an active and direct role in mobility, which is likely why the association is maintained. This is *Mixotricha paradoxa*, one of the flagellates that populate the digestive tract of termites, and its spirochetes, long helical bacteria. Discovered in the 1930s, this association was initially interpreted wrongly because the bacteria were thought to be cilia covering the flagellate. In 1964, Cleveland and Grimstone used the electron microscope to show that these "cilia" were in fact bacteria [CLE 64]. Although it has four anterior flagella, *Mixotricha* uses them not to move, but as a rudder. Movement is achieved by the coordinated undulation of the thousands of spirochetes covering its outer surface. Each bacterium binds itself by one of its extremities to a complex internal structure made of an as yet poorly defined fibrous material attached to the cytoskeleton. These new structures are unique to *Mixotricha* and are oriented towards the rear so that the attached bacterium propels its host forwards. Each bacterium has periplasmic flagella located between its internal and external membranes, which are responsible for its helical form as well as for the undulations. Molecular analyses have recently shown that there are, in fact, three different species of spirochete distributed around different areas of the host's surface, for reasons we have not yet discovered. Furthermore, we do not yet know how *Mixotricha paradoxa* and the spirochetes communicate and coordinate their movement. The spirochetes may synchronize automatically for purely physical reasons related to the flow of fluids, and the simple action of the protist's flagella may be enough to orient the direction of the movement [WEN 03]. Although this type of interaction remains rare, at least one other example is known, again in a flagellate associated with termites. In this case, 2,000–3,000 bacteria, this time in the shape of small rods, cover the surface. They use their bacterial flagella to propel the host [HON 07].

3.3. Reproduction

Some symbionts have a direct role in the reproduction of their hosts. In many animals, complex biofilms of microorganisms cover the surface of the eggs and their presence is correlated with a better hatching rate. Examples include salamanders, earthworms and some insects. These communities produce antimicrobial or antifungal substances and thus actively help prevent the establishment of detrimental microorganisms during these particularly vulnerable phases. Nonetheless, among the symbionts associated with reproductive function, the most studied are bacteria belonging to the *Wolbachia* genus. They are among the Rickettsiales, a group that includes many intracellular symbionts, including some that cause disease. *Wolbachia* lives inside the cells of its host, which is usually a terrestrial arthropod or nematode, and is transmitted either through the cytoplasm of egg cells or horizontally between species (Figure 3.4).

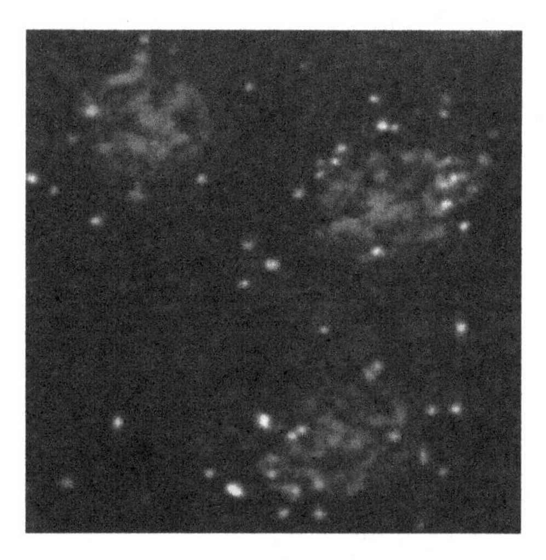

Figure 3.4. *In situ hybridization using fluorescent probes, showing early Drosophila (fruit fly) embryos in red, roughly 10 µm diameter, surrounded by Wolbachia bacteria in green. Photo reproduced with the permission of B. Loppin. For a color version of this figure, see www.iste.co.uk/duperron/symbioses.zip*

It manipulates its host's reproduction in at least four different ways [WER 97]. *Wolbachia* can induce parthenogenesis in females, i.e. the production of descendants from unfertilized eggs. This is the case in haplodiploid host species (e.g. bees), whose unfertilized eggs normally develop into haploid male individuals, i.e. with a single set of chromosomes, whereas fertilized eggs produce diploid females. *Wolbachia* causes chromosomes to double in unfertilized eggs by manipulating meiosis or mitosis. In both cases, the chromosomal number of the unfertilized eggs doubles, producing diploid females instead of haploid males. A second strategy consists of killing the male descendants (which are not likely to transmit *Wolbachia* themselves) of females infected during their early development, notably in coleoptera, diptera and lepidoptera. In a third case, *Wolbachia* can feminize genetically male individuals, particularly in woodlice, which are terrestrial isopods. In this case, the development of the gland that produces masculinizing hormones is inhibited and a female that can transmit *Wolbachia* develops "by default" [MER 09]. Finally, *Wolbachia* can induce cytoplasmic incompatibility phenomena: a non-carrier female fertilized by a carrier male will not produce descendants. A carrier male can therefore fertilize only carrier females. The mechanism is related to anomalies during the fusion of sperm nuclei and eggs and is based on a "poison-antidote" system whose determining factors are still poorly understood. These four strategies enable *Wolbachia* to spread through the host population. It is transmitted mainly through egg cells. In the first case, *Wolbachia* accelerates the production of female descendants by carrier females. In the second, the death of infected males boosts the numbers of female descendants. In the third, the bacterium transforms genetic males (non-transmitting) into females, functional or otherwise. These three strategies lead to an increase in the proportion of females and therefore an alteration of the sex ratio in the carrier population. In the final case, the bacterium limits the reproduction of non-carrier females, which in effect confers a selective advantage on carrier females as soon as the proportion of individuals infected by *Wolbachia* increases in the population. The nature of the relationship between *Wolbachia* and its hosts has been much discussed. The bacterium's ability to interfere with its host's cellular division enables it to manipulate reproduction. Initially, *Wolbachia* may therefore be considered a parasite. Nonetheless, in a population in which the bacterium has become very common, its host ultimately benefits from the infection, because its fertility is increased in relation to that of uninfected individuals. We can thus speak of benefits related to infection frequency. Furthermore, examples of induced benefits have recently been discovered, such as better resistance to RNA viruses in mosquitoes and fruit flies, resistance to insecticides, and even the production of riboflavin (vitamin B2) [HED 08, MOR 15]. In any case, *Wolbachia*'s propagation strategy is highly effective, since these bacteria are present in the cells of many arthropods, including an estimated 20–65% of insects [WER 08]. It also exists in some nematodes, which cannot reproduce without the bacterium. Indeed, in filaria, nematodes that are often parasitic, *Wolbachia* has become an obligate symbiont whose elimination leads to a decrease in mobility, reproduction and viability, although the details are not yet known. There is also competition between lineages of *Wolbachia*, resulting in complex incompatibility patterns between individuals infected by different lineages. Thus, populations of the same host species in which different *Wolbachia* are established are no longer capable of hybridizing. This reproductive isolation has been identified as a factor favoring the emergence of new host species.

Reproductive behavior is also sometimes influenced by symbionts in another way. Pheromones, which are volatile odorous substances, have many functions in animals, including attracting partners for reproduction. Decryption of the role of microbial communities in the production and alteration of these compounds has been carried out for a few years now. In crickets, for example, bacteria in the digestive tract, including *Pantoea agglomerans*, produce guaiacol and phenol, two components of the cohesion pheromone responsible for the clustering of individuals in swarms [DIL 02]. In mice, the composition of microbial communities associated with urine scent marking depends on the host's genetic factors. These communities alter the odor of the urine, notably through fermentation mechanisms, thus contributing to the preferential identification of genetically diverse partners for reproduction [LAN 07]. Studies into various mammals are ongoing, examining the role of microbial communities located in the scent glands in the differentiation of these chemical signals, depending on the individual and its affiliation to a group or genotype. These communities probably influence various social relationships, including reproduction [THE 12]. In the case of the bluebird, a North American passerine, females tend to prefer males with more colorful, bright plumage. This color difference is, in fact, correlated with a higher load of bacteria breaking down keratin on the surface of the feathers [ARC 11]. These examples illustrate the role of microorganisms on different levels of animal reproduction. The study of the influence of symbionts on reproduction and more generally on behavior is still in its infancy, due to difficulties in separating it from other possible causes, but it offers promising avenues for the future.

3.4. Protection and defense

In nature, many symbioses help protect the holobiont. Protection against hostile external factors is both the best-documented role and an expanding field of research. Protection can intervene at different levels: protection against stress related to the physico-chemical properties of the environment, against predators, parasites and even pathogens. Protection may be of a physical, chemical or behavioral nature. The details remain mostly unknown and the effects are often manifold, since one association may protect against both biotic and abiotic factors. The classification below may therefore seem a little artificial and is used only to structure the remarks.

3.4.1. *Resisting physico-chemical stress*

Throughout their lives, organisms are confronted with environments whose physico-chemical conditions are often variable and sometimes dangerous. Lichens, for example, produce a wide range of secondary metabolites, including shields against UV radiation to deal with periods of strong light, and hydrophobic compounds, which help retain water during droughts. In terrestrial plants, recent studies performed on grasses have shown that they can acquire better resistance to strong concentrations of salt in the soil and even to high temperatures through associations with particular lineages of endophytic fungi. These fungi are different from those that form mycorrhizae. Their mycelium spreads in the plant between its cells, but does not form extensions towards the outside world, unlike mycorrhizal fungi. Lineages of endophytes are peculiar to each habitat and improve the survival or biomass of the plants that contain them. Endophytes that live in plants from the edges of hot springs, where the soil is very hot, confer on them a greater tolerance to heat. For example, the fungus *Curvularia protuberata* and the geothermal plant *Dicanthelium laniginosum* can only withstand temperatures over 40 °C. By associating, the holobiont can withstand temperatures of up to 65 °C. Endophytes in coastal plants improve their tolerance to salt and those present in plants growing on agricultural land improve their hosts' tolerance to various diseases. All of these endophytes also improve the plants' resistance to drought by diminishing their water consumption by 30–50%. Rusty Rodriguez's group in the United States isolated these lineages of fungi and inoculated other plants distantly related to their natural hosts with them, showing that each tolerance could be transmitted to another host plant [ROD 08]. Interestingly, some of the fungus species involved also have highly pathogenic lineages, their positive or negative role often depending on the plants that they colonize. This illustrates the fine line between positive and negative interaction. In any case, although we do not yet know what mechanisms they are based on, these tolerances conferred by lines of symbionts peculiar to a particular habitat pave the way for important practical applications.

Although the resistance to stress conferred by symbionts is less well known in animals, the presence of *Serratia symbiotica*, a secondary facultative symbiont of aphids, seems to improve its host's survival during thermal shocks [BRO 09]. We have already discussed the role of chemosynthetic bacteria in the nutrition of many invertebrates living in sulfur-rich ecosystems, such as hydrothermal vents and mangroves. This nutrition relies on the oxidation of reduced compounds, including hydrogen sulfide (H_2S), which is a violent poison. By binding itself to animal hemoglobin and to the respiratory chain of mitochondria instead of oxygen, it causes cell asphyxiation and ultimately death. By oxidizing this sulfur to fix the carbon, the symbionts diminish its concentration in the animal cells they live in, limiting its detrimental effects. Protection through detoxification is therefore a by-product of the association. It could be called a secondary benefit or effect, since it is still not known which of the two roles, nutrition or protection, is at the root of the selection of this association. A better-known example concerns resistance to UV radiation in corals and other animals associated with phototrophic symbionts. Corals produce fluorescent pigments called pocilloporins, which both protect the holobiont by filtering harmful UV rays and absorb incident light, re-emitting it as efficient light, which can excite the chlorophyll contained in the symbionts and thus boost photosynthesis [DOV 01]. Other metabolites produced by some symbionts, called MAAs or "mycosporine-like amino acids", gather in the host's ectoderm, where they filter UV rays and limit oxidative stress. It is likely that the host itself modifies them.

3.4.2. *Avoiding predation*

Plants, like some animals (sponges, ascidians, corals, etc.), are immobile. Since flight is impossible, many of them use molecules to protect themselves from predators. Toxic or simply unpleasant, they help repel them. This molecular arsenal contains some molecules that come from symbiotic associations, produced directly by a symbiont or co-manufactured by several partners. Some endophytic fungi, such as *Epichloë*, therefore produce toxins that affect the insects that prey on their host plant [ROD 09]. This property may be attributed to the fact that these fungi derive from insect pathogens and were probably initially transmitted to plants by insects. Over the course of evolution, fungi and plants may have adapted to one another, and toxin production was maintained through positive selection due to its effect on phytophagous insects. These models informed Keith Clay's 1988 definition of "defensive mutualism", which involves at least one host, one symbiont and one enemy [CLA 14]. In the Poaceae grass *Achnatherum robustum*, nicknamed "sleepygrass" and common in the western United States, an endophyte (genus *Neotyphodium*) produces a toxic alkaloid from the ergoline family (of which a well-known example is rye ergot), d-lysergic acid amide, which is capable of putting a horse to sleep for several days, even in low doses. Mammals avoid this type of plant. Lichens also produce substances that repel herbivores. As for animals, most predators reject the larvae of the bryozoan *Bugula neritina*, which are covered in bryostatins, molecules that can act as repellents. These molecules are produced by the bacterium “*Candidatus* Endobugula sertula”. Bryostatins are much less abundant in the adult, which is protected from predators by a chitinous exoskeleton and by its colonial way of life. In some soft and gorgonian corals, the production of diterpenes, which make the coral inedible to predators, has been attributed to symbionts of the *Symbiodinium* genus, which also ensure their symbiotic nutrition [LOP 14].

3.4.3. *Fighting parasites and pathogens*

Parasites and pathogens bind themselves to their target or penetrate its tissues. The simplest way for symbionts to confer resistance to their hosts is therefore to prevent the entry of parasites and pathogens into the organism by occupying the exposed outer surfaces. Competitive exclusion relationships are established between species, in which resident symbionts often produce antibacterial, antiviral or antifungal substances. This is the case in communities that inhabit the digestive tract of animals, in which the presence of the gut microbiota physically leaves little space for pathogens to establish themselves. A similar phenomenon is observed on the skin of amphibians such as frogs and salamanders, on whose surface a bacterial community produces antifungal substances, such as violacein. This prevents fungal attacks, such as chytridiomycosis, a fatal infection that is contributing to the current decline of amphibian populations across the world [LAU 08, HAR 09].

Another strategy is to stimulate the host’s defenses, which many symbionts of insects and some endophytic fungi do. Their presence alone is correlated with a faster response from the host during infection. In plants that possess certain endophytes, for example, an increased production of lignin is observed, which forms a physical barrier preventing infection by other microorganisms [RED 99]. The presence of mycorrhizae also stimulates production of lignin and tannins by the root, while the fungus itself produces antibiotics. Lichens produce massive quantities of antibacterial and antifungal molecules [HUN 99]. Moreover, lichens are, once again, particularly well equipped by their secondary metabolism, which is probably related to their exceptional longevity. Although most of their compounds come from the fungus, the fungus only produces them when associated with its photosynthetic partner, suggesting that the presence of the phycobiont is, in one way or another, necessary.

If, despite all of this, the parasite or pathogen manages to penetrate the host organism, the symbionts can contribute through another line of defense. In aphids, the bacterium *Hamiltonella defensa*, a secondary facultative symbiont, protects the host from the parasitoid wasp *Aphidius ervi* [OLI 08]. The wasp lays eggs directly inside the aphid and its larva develops by devouring it from within. The protection conferred by *Hamiltonella* relies on the bacterium's production of toxins that are fatal to the *Aphidius* larva but not to the host aphid. Unexpectedly, these toxins are themselves coded by a bacteriophage (i.e. a virus that infects bacteria) whose genome has established itself within that of *Hamiltonella* [MOR 05].

Details of protection mechanisms are often unknown and particularly difficult to break down, as is the case for the aforementioned resistance to viruses observed in mosquitoes and flies infected by *Wolbachia* [HED 08]. The mere notion of protection may be misleading, because protection of a host organism often consists of attacking its antagonist. Indeed, some of these associations may be seen as defensive symbioses from one angle and offensive associations from another; it is a question of point of view. Protective symbioses are one of the most promising fields of study, offering great prospects for the development of strategies for the protection and biological improvement of species and cultures and even for the fight against epidemics.

4

Outline of How Symbioses Work

Symbiotic associations may have many roles, including those described previously, which are a few of the best-documented examples. Many others are yet to be discovered. We may, therefore, wonder whether these various symbioses share any common traits to justify the concept. From the point of view of interaction between partners, symbiosis can be broken down into several stages modeled on the system's life cycle. The interaction begins with the symbiosis acquisition stage, when the partners meet. Then comes initiation and continued dialogue throughout the interaction. It may also end at certain moments, for various reasons. Each stage requires the installation of dialogue mechanisms to ensure that potential conflicts between partners can be surmounted. As we will see, in this respect, it is interesting to consider the host – by convention the larger partner – as both an organism and an ecosystem, in which symbionts must find their niche.

4.1. Symbiosis acquisition

4.1.1. *Horizontal transmission*

A symbiotic association can be maintained in several ways. The most common involves reestablishing the association in each new host generation, through an encounter between an aposymbiotic host (lacking symbionts) and its symbionts [BRI 10]. The host can therefore exist autonomously during all or part of its own life cycle and does not receive its symbionts from its parents. This symbiosis establishment mechanism is called horizontal transmission (Figure 4.1). From the symbiont's point of view, there are two methods. The symbiont may pass from one host to another by passing briefly through the environment, sometimes unable to spend its entire life cycle there. This is lateral transmission. Alternatively, the symbiont can come from a population of free-living forms that live in the environment, rather than inside any host. This is environmental acquisition. Symbiosis is usually facultative only for the symbionts; this is rarely so for hosts, for whom the aposymbiotic state is often only transitory.

Interaction between future partners begins well before contact, with detection and reciprocal recognition in the environment, assuming each partner to be in a favorable phase. In plants, nodulation and establishment of mycorrhizae occurs continuously as the roots grow and communicate with the free-living forms of nitrogen-fixing bacteria or mycorrhizal fungi hyphae present in the soil. In principle, the digestive tract of animals may also be continuously colonized by new types of symbionts present in food or acquired through interaction with other organisms. In reality however, most of the symbionts will establish early in life. Young xylophagous termites acquire their initial working digestive flora through social interactions with other members of the colony, by coprophagy (eating feces) or proctodeal trophallaxis (anus-to-mouth feeding). They reacquire symbionts after each molt by the same mechanisms. The first human intestinal flora are acquired during childbirth and breastfeeding. Both are examples of horizontal transmission, where symbionts are not necessarily (but can be) obtained from the parents, although the microorganisms have not necessarily existed in a free state in the environment. Most benthic marine invertebrates have a relatively short "competence window" for acquiring symbionts. This is usually when the larva, a dispersive, often planktonic, stage, establishes itself on the definitive substrate where it will metamorphose to the adult stage. This corresponds to a lifestyle change and the establishment of the symbiosis is often crucial for survival. Initially, colonization by symbionts is dense and not very specific, but when the window closes, stringency increases and bacteria are tolerated only in their destination tissues. This is how bivalves and chemosynthetic annelids, whose larval stages are aposymbiotic, such as *Bathymodiolus* and *Riftia*, acquire their bacteria soon after installation on the seabed. This involves pumping large quantities of water in order to maximize the chances of encountering free-living forms of their symbiotic bacteria present in the environment. Infection is initially general in many tissues (surface of the epithelia in *Bathymodiolus*, tegument and connective tissue in *Riftia*), and then limited to the gill and trophosome, respectively. In corals which acquire their symbiotic algae from the environment, the installation of the symbiont itself induces the larva's fixation to the seabed, its metamorphosis and the start of calcification. Adults often still have a limited capacity for the acquisition of new bacteria; nonetheless, if initial infection does not occur, the host usually does not survive [BRI 10].

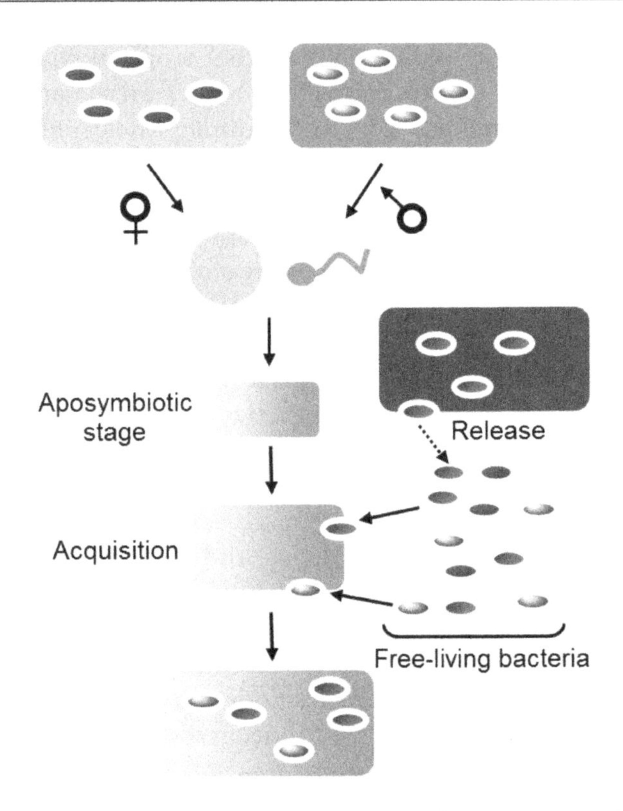

Figure 4.1. *Horizontal transmission of symbionts from one generation to the next. Neither the gametes nor the zygote created by their fusion contains symbionts. It experiences an aposymbiotic phase of varying length. Symbionts are acquired from other hosts (lateral transmission) or from the environment, in which free-living forms of the symbionts exist (environmental acquisition). These free-living forms may be rare or abundant, active or dormant, and are often little known. They may have been released or liberated into the environment by individuals, parents or otherwise. Acquisition may be limited to a particular phase of the cycle (competence window, as in marine invertebrates) or continue throughout the life cycle (some digestive tract bacteria). The symbiont lineages present in an individual may therefore differ from those present in its parents. The diagram depicts a theoretical example in which lateral transmission (through release, green bacteria on the diagram) and environmental acquisition (by free-living forms, yellow bacteria) coexist. For a color version of this figure, see www.iste.co.uk/duperron/symbioses.zip*

Recognition between host and symbiont involves diverse chemical compounds. Mychorrhizal fungi appear to be able to recognize their host plant using strigolactones produced by their roots. However, one of the best-known examples is nitrogen-fixing bacteria. The free-living forms of most Rhizobia have a flagellum and are therefore mobile; they move towards the host plant's root through chemoattraction, i.e. they are attracted to substances emitted by the roots, notably flavonoids, which also stimulate bacterial growth. Host and symbionts therefore recognize one another and bond through the interaction of proteins located on the surface of the bacterium and recognized by the plant cell receptors, and by plant lectins, which interact with bacterial exopolysaccharides. In response to the perception of plant signals, the bacteria produce proteins called nodulation (Nod) factors. These factors interact with the plant's receptors; this interaction causes a cascade of modifications. These include the tip of the root hairs curling around the bacteria. This deformation is due to the activation of the plant's ENOD (early nodulin) genes, which ultimately lead to nodule formation. The bacteria then move into the root along an infection thread, a tunnel generated by the apoptosis (programmed cell death) of plant cells. The bacteria are then internalized by the root cortex cells, which proliferate and form root nodules, bulges within which bacteria can multiply and fix atmospheric nitrogen. Later, the plant cells produce leghemoglobin inside the nodule.

In animals, symbiosis establishment mechanisms are documented in the squid *Euprymna scolopes*. Bioluminescent *Vibrio fischeri* bacteria represent only 0.1% of the bacterioplankton. Gram-negative planktonic bacteria are captured if they approach the special mucus-coated ciliate epithelium that lines the light organ appendages [MCF 14b]. They are then gathered in this mucus. *V. fischeri* is the only species that can migrate through the pores and reach the crypts (Figure 4.2). Thanks to its pili and to some host's proteins that bind to its membrane lipopolysaccharides, *V. fischeri* adheres and enters through the pores into a canal, which directs it to one of the symmetrical lobes that make up the light organ. *V. fischeri* is transported through a canal through ciliary beating. It reaches the crypts by following a food source gradient, in which the symbiont's presence stimulates production. All other bacteria are gradually eliminated during the journey, probably through the action of the mucus, which may act as an anti-microbial cocktail, with the production of peptides, toxic nitrogen species and an oxidative stress. *V. fischeri* is able to survive these treatments. Symbiont catalase, for example, traps the hydrogen peroxide from which reactive oxygen species are produced, limiting the detrimental effects of the oxidative stress generated by the host's cells. Ultimately, only one or two bacteria will colonize each crypt and induce the multiplication of microvilli, tissue folds on the surface from which they establish themselves [NYH 04]. Bacteria proliferate in these crypts (Figure 4.2), but each morning 90% of them are discharged, turning off the light signal. Those that remain begin to grow again. After the initial colonization phase, no other bacteria will colonize the crypts. Indeed, the presence of *V. fischeri* in the crypts induces the apoptosis of the surface ciliate epithelium and the narrowing of the canals through which it entered. In the annelid *Riftia pachyptila*, symbionts penetrate the epiderm, cross the mesoderm and reach the trophosome in formation [NUS 06]. This involves crossing through the cells encountered, but the mechanisms are not yet understood.

Horizontal transmission therefore involves complex mechanisms, mainly including phenomena of attraction by specific chemical compounds, enabling the encounter between partners, and protein–sugar recognition in the cell membranes, enabling adhesion, circumvention of immune responses and the colonization of the target organ. One classic response of the host cells is the generation of an oxidative stress, to which the symbionts have usually become resistant [MON 14].

Figure 4.2. *Seen under a fluorescence microscope, two lobes of the ventral light organ in a young Euprymna scolopes. The pores through which Vibrio fischeri enters are visible on each side (one indicated with a white arrow). The crypts of the lobe on the left of the image have already been colonized by bacteria (green), while the lobe on the right has not. Photo reproduced with permission of M. McFall-Ngai. For a color version of this figure, see www.iste.co.uk/duperron/symbioses.zip*

4.1.2. *Vertical transmission*

Vertical transmission of symbionts assumes that the association is permanent and that there is no aposymbiotic phase for the host, nor sometimes a free phase for the symbiont (Figure 4.3). Symbionts usually pass from one generation to the next by positioning themselves inside the egg cell (intracellular symbionts) or on its surface (extracellular symbionts). Examples of the first case are *Buchnera aphidicola*, the main symbiont of the aphid, which passes from the mother's bacteriome to her ovocytes, some corals, which practice vertical transmission during sexual reproduction, and even the chemosynthetic symbiont of vesicomyid marine bivalves such as *Calyptogena*, which mainly colonizes the gill but is also present in the gonads and egg cells at all stages of ovogenesis [END 90, KOG 12]. The tsetse fly is viviparous and has special glands that produce a liquid containing the symbiont *Wigglesworthia*, with which it feeds and transmits symbionts to its embryos. In these various cases, transmission is thus strictly maternal. Biparental transmission is, however, sometimes possible: one example is the secondary defensive symbiont of the aphid, *Hamiltonella defensa*, in which transfer from the males during mating results in co-infections in the females and therefore the transmission to their descendants of one or other of the parental lines present [MOR 06]. In organisms that practice asexual reproduction, such as corals, cells that differentiate into new individuals contain symbionts from the parent organism, which guarantees the continuation of the association in the newly formed individual. This is also the case in lichens. Although the sexual reproduction of lichen-forming fungi does not transmit phycobionts, some form dual propagules for asexual reproduction, called soredia or isidia, in which the fungal hypha surrounds the cells of the photosynthetic partner to ensure the co-dispersion of the two partners.

The transfer of symbionts from the usual organs where their functions occur to the reproductive cells can take multiple forms and is often poorly understood. In aphids, however, this journey is well established. Bacteriocytes are found in the hemocoel, the general cavity that contains all organs, including the ovaries. Bacteria leave these bacteriocytes through exocytosis and cross the space between the bacteriocyte and the embryo. Up to many thousands of them penetrate the embryo's cells at the blastula stage, through endocytosis [KOG 12]. In another sense, the processes that enable the sequestration of symbionts initially present in the egg cell in a particular organ during early development are also little understood. During the early embryonic development of the aphid, a specific cell lineage differentiates into bacteriocytes before symbiotic infection has even occurred, suggesting that this determination process depends entirely on the host, at least until this stage [BRA 03]. The molecular mechanisms at play are, however, still unknown.

Generally speaking, transmission is rarely strictly vertical and various parallel mechanisms often add an element of horizontal transmission whose evolutionary consequences can be very significant, since it enables symbiont shuffling. In lichens, for example, sexual reproduction of the fungus does not transmit the phycobiont, whereas the phycobiont and mycobiont remain physically associated in the same propagule (soredia or isidia, Figure 4.4) during asexual reproduction. Horizontal and vertical transmission ultimately share few common characteristics, besides both being able to bring about obligate associations for at least one partner and being efficient when it comes to the survival of the holobiont.

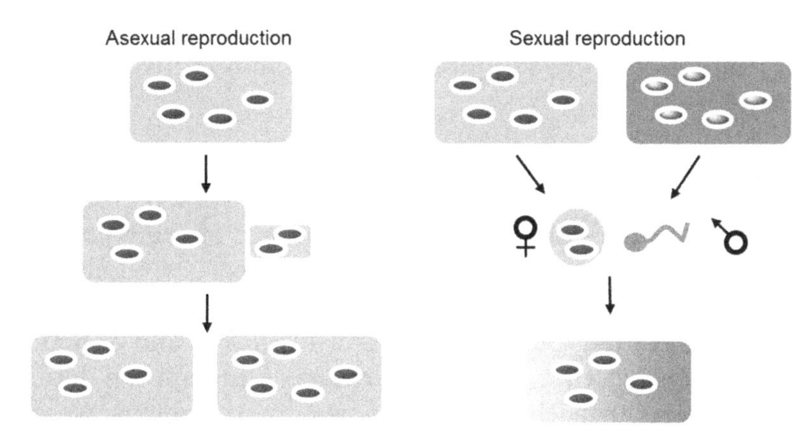

Figure 4.3. *Vertical transmission of symbionts from one generation to the next. Left, transmission in clonal asexual reproduction. A new individual can be seen differentiating from a region of the parent. It contains a limited number of symbionts from the parent organism. It can then become autonomous, as observed in lichens (Figure 4.4). Right, sexual reproduction. The parents produce gametes, here a large female gamete containing a few symbionts from the mother (left) and a small mobile male gamete containing no symbionts. The zygote, the individual resulting from the fusion of the gametes, will have mixed genetics from the two parents, but all of its symbionts will come from the mother. This is strictly maternal transmission. Cases of co-transmission by both parents, in which both gametes contain symbionts and the descendants inherit symbionts from both parents, are rare. In all cases, reproduction transmits only a very small fraction of the symbionts, introducing a bottleneck into their life cycle. For a color version of this figure, see www.iste.co.uk/duperron/symbioses.zip*

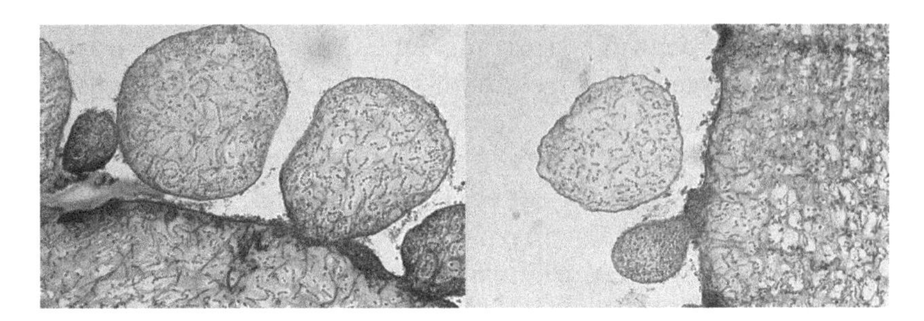

Figure 4.4. *Lichen isidia. In some lichens, small outgrowths form on the surface of the thallus. They contain both the hypha of the mycobiont (purple) and the cells of the phycobiont (green). By detaching themselves, they enable the fungus to reproduce asexually, ensuring the transmission of the photosynthetic partner. For a color version of this figure, see www.iste.co.uk/duperron/symbioses.zip*

4.2. Dialogue between host and symbionts: who is in charge?

Symbiosis is an exchange of good procedures. For it to work properly, there must be uninterrupted dialogue between partners and constant adjustment of their position and respective contribution, all in the context of a potentially changeable environment. This requires the association to be controlled. Either party may exercise control and sanction mechanisms exist for cheating partners. Although the details of the molecular mechanisms often remain mysterious, we are at least beginning to realize how precisely the associations are regulated.

4.2.1. *Adjustment of the symbiosis with respect to the environment*

Deep-sea hydrothermal vents are particularly temporally and spatially variable environments. The fluid output flow can change in a few minutes or even seconds, so chemosynthetic animals living nearby are exposed to considerable variations in temperature, oxygen content and even energy sources for their symbionts. In the case of *Bathymodiolus azoricus* mussels, symbiosis involves two types of bacteria, one using reduced sulfur and the other methane, two compounds that also vary in abundance. Having two distinct metabolisms to acquire its carbon is a considerable advantage for the mussel in these changeable environments. *B. azoricus* colonizes various hydrothermal sites along the Mid-Atlantic Ridge. The relative abundance of its sulfur-oxidizing and methanotrophic symbionts, like their contribution to their host's nutrition, correlate with the respective availability of sulfur and methane on each site. We therefore observe the predominance of sulfur-oxidizing bacteria on sites that are rich in sulfur and low in methane, and of methanotrophs on sites that are rich in methane and lower in sulfur [DUP 10]. Seeking to understand the dynamics of such an association, studies performed by moving mussels from sites that are rich in reduced compounds to sites that are less active have shown that the abundance of symbionts decreased when they were moved further from their energy sources. Using *in vivo* experiments performed at atmospheric pressure or in a hyperbaric chamber (to recreate the high pressure of the seabed), various studies have shown that the relative abundance of sulfur-oxidizing and methanotrophic bacteria varied in only a few hours in the gills, in direct correlation with the availability of sources. If sulfur is added, the former take over; if methane is added, the latter take over. This example shows that, in an association involving several types of symbionts, there are mechanisms enabling the position of each one in relation to environmental parameters to be adjusted and that this adjustment can be very rapid. Multiplicity of partners and flexibility enable acclimatization to variable environments and are among the characteristics that contribute to the success of the mussels and of many other types of holobiont in nature. In the case of the mussels, we do not know if this results from active control by the host, direct competition between symbionts or a mixture of the two. However, in other models, the dialogue between host and symbionts is beginning to be better understood and we are starting to understand how each partner ensures that the other plays its part.

4.2.2. *Selective cooperation and the adjustment of contributions*

The mycelium of a mycorrhizal fungus can associate simultaneously with many plants, potentially from different species, to which they provide nutrients. Furthermore, a single plant is itself associated with many lineages of mycorrhizal fungi, the whole forming an interconnected network. A recent study has shown that a mycelium is able to discriminate between its various plant partners. Of the plants to which it is connected, the mycelium provides more phosphorous and nitrogen to those that are the best-exposed to light, i.e. those that provide more carbonated substrates from photosynthesis. At the same time, however, it continues to provide nutrients to shaded plants, even though they "pay" less in return [FEL 14]. Besides, root colonization of shaded plants is more efficient if well-exposed plants are also associated with the mycelium. This suggests that part of the benefit drawn from exposed plants is reinvested by fungi in the upkeep of the association with shaded plants. Such a strategy is evidently wise, since the exposure of various plants to light is likely to vary over time. A comparable mechanism exists in plants, which specifically reward their best mycorrhizal partners, i.e. those that provide more phosphorous, by allocating them a larger portion of carbonated derivatives [KIE 11]. An increase in the transport of carbonated derivatives to the fungus stimulates the fixation and transfer of phosphorous in the latter, shown by the correlation between levels of expression of proteins transporting carbon and phosphorous between the partners. All of this proves the existence of a refined and individualized dialogue between the partners, each able to measure what it gives and receives from each of the others. This fine regulation of the interaction, adjusted for each partner in the event of multiple symbioses, limits the emergence of cheating strategies on both sides, which can cause one partner to enslave the other. In a very complex network associating multiple mycorrhizal fungi with multiple plants, this type of mechanism is probably necessary for the upkeep of the association and likely has a positive effect on the whole ecosystem.

4.2.3. *Maintaining cooperation by sanctioning cheaters*

In some cases, one partner is clearly in control of the situation. This is the case in associations in which the symbionts are located within the host, even if their transmission is environmental. Through major studies performed on the symbiosis between nitrogen-fixing legumes and *Rhizobium*, Toby Kiers's group has artificially reconstructed a cheating situation by replacing the dinitrogen with argon gas in the atmosphere around some of the nodules. The bacteria found there can no longer fix nitrogen, nor, therefore, transmit it to the host plant, simulating the association with a non-cooperative lineage [KIE 03]. In theory, the bacteria could reinvest the economized energy in their own growth and thus supplant the nitrogen-fixing lineages. The opposite has been observed, however: nodules containing the "cheater" lines were smaller and liberated fewer, less viable bacteria into the environment than the nitrogen-fixing nodules (Figure 4.5). This is due to the plant's lesser investment in these nodules, since its response to the "cheat" is to provide less oxygen to the nodule, slowing the metabolism of the symbionts. The level of sanction was even shown to be proportionate to the reduction in nitrogen fixation [KIE 06].

A similar mechanism may be at play in the association between the squid and *Vibrio fischeri*. It has been shown that deficient mutants of *V. fischeri* for one of the genes responsible for light production cannot be experimentally maintained in the light organ [MCF 14a]. These genes code for either luciferase, a catalyst for light production, or the quorum perception mechanism, which tells the population when to emit it (mechanism shown in Figure 3.2). The general structure and physiology of the light organ are similar to the squid's eye, including the presence of rhodopsin, a photosensitive pigment that perceives light. The host can therefore perceive the emission of light by the symbionts and, in this particular example, its absence. Although not completely understood, the disappearance of these lineages may be related to a reaction of the host.

The existence of sanction mechanisms suggests that the plant and the squid can accurately measure the output of their symbionts, which is likely to be the case. This system makes it possible to limit the emergence of cheating or profiteering partners and helps stabilize the association in the long term. This is of vital importance in the case of bacteria acquired from the environment, since cheater lineages are known to exist. For example, one study isolated in nature some nitrogen-fixing bacteria of the *Bradyrhizobium* genus, capable of nodulating and proliferating in the roots without giving anything in return [SAC 10]. Similar systems doubtless exist, therefore, in many other associations.

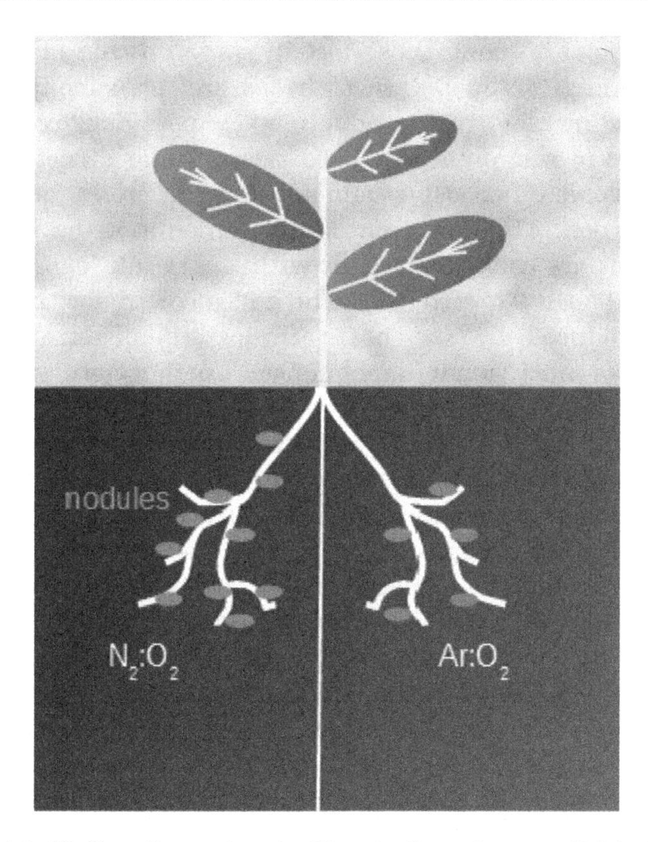

Figure 4.5. *"Split-root" experiments. The plant's roots are split into two, the left side receiving nitrogen and the right an inert gas, argon (Ar). The pink nodules contain Rhizobia. On the left, Rhizobia can fix nitrogen, but not on the right, simulating a cheating situation. The nodules will be half as numerous on the side exposed to argon. The reproductive success of the bacteria is therefore lesser when they do not render the service expected by the host, in this case nitrogen fixation. This is due to a sanction by the host, which limits the oxygen supply to the "cheating" nodules. Diagram adapted from [KIE 03]. For a color version of this figure, see www.iste.co.uk/duperron/symbioses.zip*

4.2.4. *Forced cooperation by integration of the symbiont*

When symbiosis is obligate for the symbionts and vertically transmitted, as in the example of *Buchnera aphidicola* in aphids, the situation changes radically. Such an association is accompanied by a massive loss of genes in the symbiont. Among those lost in *B. aphidicola* are all the genes coding for lipopolysaccharides and phospholipids and some coding for the production of peptidoglycanes. As explained in section 4.1.1, these extracellular structures are often involved in host–symbiont recognition in horizontal transmission. The genes encoding elements of the bacterial flagellum are also lost and with them all capacity for mobility [WER 02, MCC 12]. The genes coding for the amino acids produced by the host are lost, demonstrating the symbiont's nutritional dependence. Furthermore, the host compensates by over-expressing them to feed its bacteria. In vesicomyid bivalves, the symbiont *Vesicomyosocius okutanii* has even lost the gene coding for FtsZ, a protein responsible for the formation of the ring at the future site of the septum of cell division, suggesting that the host even controls its symbiont's division mechanism [KUW 07]. Similarly, *Wigglesworthia*, a symbiont of the tsetse fly, has lost the gene *dnaA*, which is involved in initiating the replication of its own DNA, once again suggesting control by the host [MOY 08]. The genes retained are those involved in the symbiont's expected functions, such as the provision of amino acids essential for the host in *Buchnera*, the oxidation of reduced sulfurs coupled with the Calvin cycle to fix inorganic carbon in *Vesicomyosocius* and the production of vitamins in *Wigglesworthia*. The symbionts' loss of autonomy gives the host full control of the situation. For example, the genes producing the amino acids required by the aphid are not only present but also highly expressed by *Buchnera* in response to its host's demand. Although modeling approaches have shown that this type of relationship temporally stabilizes the association, the modalities of the dialogue have therefore been reviewed in depth, with evolutionary consequences that will be discussed in the next chapter.

4.3. The end of the symbiosis: trapped, digested or discharged?

Obligate symbioses with vertical transmission, such as that of the aphid, unite the destiny of the partners, and the host's death leads directly to that of the symbionts, which cannot survive it. Effectively, therefore, this is not the end of the symbiosis as an association, but of the entire holobiont. In other cases, the end of a symbiosis can trigger many situations. In most nutritional symbioses involving chemosynthetic bacteria, the bacteria themselves are digested within the cells of the tissue of the animal that contains them, putting a *de facto* end to the symbiosis, at least at an individual bacterium's level. Of course, these losses are compensated for by the growth and division of other bacteria present in the system, or through the acquisition of new symbionts from the environment. This lasts as long as the symbionts have substrates for their metabolism, but if these disappear, the number of symbionts diminishes, ultimately leading to the host's death [KAD 05].

However, the host's death when substrates are still abundant can lead to the liberation of the symbionts and their return to a free-living state in the environment. This has been shown very recently in the hydrothermal annelid *Riftia pachyptila* [KLO 15], whose symbionts are capable of returning to free life from the debris of the trophosome. They express a flagellum, absent in symbiotic forms, which enables them to move actively, and they rapidly start dividing in the environment. In photosynthetic symbioses, such as that of corals, the symbionts release an exudate of carbonated compounds to their host and are therefore rarely digested [TIT 96]. However, a sudden change in the environment, such as an increase in water temperature, in UV rays or a strong increase in the quantity of CO_2, can cause coral bleaching. This phenomenon is due to the discharge of symbionts through exocytosis, deconstruction or degradation of animal tissues by apoptosis phenomena. The disappearance of symbionts considerably reduces nutrients input in colonies and can lead to significant mortality rates in corals. In the *Euprymna* squid, 90% of the *Vibrio fischeri* present in the crypts are expelled each morning, when the animal hides away for the day. In these three cases, for very different reasons and with very different consequences for the host, the symbiosis ends here for the microorganisms, which return to free life. This return gives the bacteria the opportunity to colonize new hosts. This aspect of symbiotic interaction and its evolutionary consequences is only just beginning to be explored.

5

Symbiosis and Evolution

Symbioses work in different ways, but some common traits can be identified. In most symbioses acquired by horizontal transmission, recognition is based on chemotaxis mechanisms, meaning that the symbiont is attracted by molecules produced by the host. Contact is based on ligand–receptor interactions. The symbiont is often faced with hostile conditions, such as the presence of antimicrobial peptides and oxidative stress, to which it may be more or less adapted. Finally, if symbionts are internalized, this occurs through endocytosis. For vertically transmitted symbioses, the path is often a bit shorter, but if the host is pluricellular, passage from the organ where the symbiont functions to the offspring may also involve exocytosis and endocytosis phenomena. The host often has dedicated structures for receiving symbionts, but their development is conditioned by interaction with the latter. Although the symbionts are a sort of "VIP" for their host, the latter must not allow itself to be taken over. Regulation of the density of symbionts and their activity often seems to be based on bacterial quorum-sensing phenomena. The host must avoid being exploited or invaded. Mechanisms are therefore implemented to control the function and number of symbionts, whose growth may be limited by digestion by the host or by adjustment of the supply of a limiting nutrient, such as nitrogen in coral.

Generalization is impossible, but this non-exhaustive list of characteristics common to a large number of symbiotic associations is proof of a key fact: these interactions of communication and physical association are not spontaneous and each stage is a product of evolution. This evolution leads to adaptation to precise ecological niches, in which organisms living in symbiosis have been retained by natural selection and are often important players. One major challenge for current research is therefore understanding the evolution of each of the various symbioses. The idea is to measure to what extent symbiosis, as a phenomenon of inter-species association, has shaped the evolution of life on our planet.

5.1. Becoming a host or a symbiont

The first question we might ask ourselves is why there are so many symbioses. Lynn Margulis was one of the major figures in the discipline from the 1970s onwards, reintroducing the idea that the eukaryotic cell had endosymbiotic origins. In an interview recounted by Douglas Zook, she explains that eukaryotic cells have a remarkable capacity for incorporating, and sometimes fusing with, other cells through endocytosis phenomena, which can be attributed to the highly dynamic behavior of the membrane [ZOO 15]. Indeed, this is a hallmark of the group. She illustrates this with two examples: sexual reproduction, which involves one gamete penetrating another and the fusion of their respective nuclei, and the intracellular digestion of microorganisms by many eukaryotes, such as paramecia, which must first ingest them. In this case, endocytosed cells are found in delineated compartments, isolated from the cytoplasm of the host cell by membranes. For Margulis, the origins of the ability to establish symbiosis lie in this propensity to allow the entry of foreign cells. Subsequently, the cell can digest the intruder, but it can also refrain from doing so, or simply delay the process until the incorporated cell has had time to grow or multiply. Broadly speaking, this is how an organism becomes a host. This view invites new interpretations. Thus, for example, in mussels living around hydrothermal vents, we might suggest the currently untested hypothesis that the gill cells allow methanotrophic or sulfur-oxidizing bacteria to enter and give them time to grow and divide before digesting them. Therefore, in this example, symbiosis seems to be a phenomenon due simply to the temporary maintenance (or delayed digestion) of these specific types of bacteria in the cells. One way of enabling the installation and maintenance of symbionts is to lower one's guard. For example, the immune system of the aphid is weak. Gene expression patterns do not change when the organism faces a bacterial attack. Indeed, many genes that are key to the detection of bacterial attacks and the immune response of insects (including effectors responsible for eliminating intruders) are absent in the aphid genome [CON 10]. This "strategy" is risky, since it weakens the organism in the face of potential pathogens. To determine exactly what enters, what is maintained and what is digested, many hosts and symbionts implement a complex dialogue.

Communication mechanisms also come into play and dictate how an organism becomes a symbiont. Historically, and for obvious reasons, the study of pathogens has preceded that of symbionts and is more advanced. However, one major observation over the last few years has been that of the striking similarities between the mechanisms used by pathogens and those used by more beneficial symbionts. Quorum sensing by bacteria was discovered in *Vibrio fischeri* and is an integral part of the dialogue with the squid, triggering the emission of light in the light organ. It is also known to influence the efficiency of nodulation and nitrogen fixation by bacteria such as *Rhizobium* [GON 03]. But many pathogens use the same systems to coordinate their activity: for example, triggering or reducing the production of substances responsible for detrimental effects only when a certain density is reached [MIL 01]. For example, *Vibrio cholerae*, the cholera agent, produces toxins when its density is low and its virulence diminishes when the bacterial density increases. In some pathogens, this strategy is interpreted as a way of delaying the immune response until the pathogen is already abundant. Some pathogens and symbionts also use similar secretion systems to transfer effector proteins to another cell. These systems, called type III and IV, specialize in the translocation of proteins from the bacterial cytoplasm to the exterior through a third membrane (beyond the bacterial inner and outer membrane), typically that of a eukaryotic cell. They are rather like syringes that help to inject proteins. Genes encoding these systems are found alongside those coding for toxins on "pathogenicity islands" in the genome of some pathogens, such as *Helicobacter pylori* and *Salmonella enterica* [SCH 04]. In many symbionts, such as the protective bacterium of the aphid, *Hamiltonella defensa*, these secretion systems have been recruited to transfer other types of protein that are not toxic to the host [DAL 06].

The sharing of common mechanisms between symbionts and pathogens shows that the line is thin. In the microbial world, these mechanisms are often acquired through lateral gene transfer, i.e. the integration of a fragment of genome from a non-related bacterium. Another source is simply common ancestry between pathogens and symbionts. This is how *Vibrio fischeri*, the bioluminescent bacterium, is related to the cholera agent *V. cholerae*, which explains the sharing of certain characteristics. This sometimes troubling proximity between pathogens and beneficial symbionts suggests that, in some cases, symbiosis may originate in a form of domestication of a pathogen, whose capacity to harm is first reduced [SAC 11]. For example, the symbionts of the annelid *Riftia pachyptila* have been observed crossing through its tissues as pathogenic bacteria would, before installing themselves in the trophosome, where their role is beneficial [NUS 06]. The opposite is also possible and there are examples of mutualism turning to parasitism in the endophytic fungi of plants [ROD 09]. Indeed, a study has shown that a simple mutation in a signaling pathway in the fungus *Epichloë festucae* can turn a mutualist symbiont into a pathogen in the association between the fungus and the ryegrass *Lolium perenne* [EAT 11].

"Pathogenicity islands", coherent groups of genes responsible for the virulence of some pathogens, are often acquired through lateral transfer. The existence of "symbiosis islands" is questionable. For example, this analogy may be used to describe the plasmid or genomic island (depending on the species), which includes genes for nodulation, nitrogen fixation and the biosynthesis of polysaccharides in some nitrogen-fixing bacteria [GON 03]. In this precise case, it is thought that the physical grouping of these genes enabled both the appearance of nodulation around 65 million years ago and the capacity for nodulation to spread through lateral transfer in many groups of only moderately related bacteria [REM 16]. In some cases, the modification of a single gene can be enough to transform a neutral bacterium into a symbiont.

5.2. Maintaining a symbiosis

Maintaining a symbiosis from one generation to the next is not straightforward. The previous chapter discussed the complexity of the mechanisms in play in the *de novo* establishment of the association or its transfer to each new generation. These mechanisms are costly and subject to natural selection. This time, it is the holobiont that must be selected. In many of the cases discussed so far, the new function acquired through symbiosis is a selective advantage in itself. The capacity to assimilate carbon or nitrogen from atmospheric gases, to better exploit the nutrients that are available or to produce those that are absent in food, to improve reproductive performance and to acquire protection are of obvious benefit. This type of evident benefits is behind the classic definitions of symbiosis. However, the story is not always so simple, and in some cases the reason for the initial selection of a symbiosis may be very different to the reason for its success today. The most classic example is the mitochondria of eukaryotic cells. Today, a mitochondrion is a factory where energy is produced from the aerobic oxidation of organic compounds. The Krebs cycle, which is particularly effective at this, occurs there. Thanks to mitochondria, the eukaryotic cell obtains much more energy per carbon unit than a prokaryote could, which has probably enabled eukaryotes to reach greater sizes [MCF 15]. Mitochondria derive from a symbiotic bacterium [BAU 14]. However, the eukaryotic cell originated between 1.6 and 2 billion years, in an epoch when oxygen was still rare in the atmosphere, so its production of aerobic energy was probably not the reason it was initially adopted. The debate into the original nature of the relationship is ongoing. A role in oxygen detoxification has long been suggested, but this hypothesis is increasingly called into question. More recent hypotheses suggest syntrophy with the proto-eukaryote, i.e. the production of substrates for its metabolism, including dihydrogen, dihydrogen sulfide, ATP and organic acids. In any case, the massive production of ATP, which is behind the success of current eukaryotes, seems to have been a favorable secondary effect, rather than the initial reason for the selection of the association. This brings us to one of the mechanisms that can contribute to the maintenance of a symbiotic association: pre-adaptation or exaptation. This is the idea that an association does not necessarily make the organism more effective at the time, but enables it later to adapt to an as-yet-unencountered new environment. This is the case for mussels living around deep-sea hydrothermal vents. There is now a consensus that symbiosis with sulfur-oxidizing bacteria appeared well before the colonization of these sources and that it initially enabled the mussels to live in shallow, sulfur-rich habitats, such as whale carcasses and sunken wood, in which the massive production of sulfurs is the result of the decomposition of organic matter by microorganisms in anaerobic conditions. Due to the toxicity of the sulfurs for animals, the competition was fairly low there and the mussels diversified. Then, during evolution, some lineages colonized the hydrothermal vents and cold seep sites, which are also sulfur-rich. They were pre-adapted there due to their symbiosis and found it a particularly favorable habitat, in which they could reach previously unseen sizes and densities. In this case, the possibility of colonizing new ecological niches contributed towards maintaining the association. As the association becomes more and more integrated, its maintenance is then facilitated by the implementation of transmission strategies, until the vertical transmission phase, where maintenance of the association is guaranteed over a long period, although symbionts may still be replaced.

5.3. Coevolution, co-speciation and asymmetry in the symbiotic relationship

Coevolution is a phenomenon in which the evolution of one organism influences that of one or several others. Co-speciation is a specific case in which a speciation event in an organism – when one evolutionary lineage produces two lineages – is accompanied by a parallel speciation event in another (Figure 5.1). Co-speciation may be a consequence of coevolution, for example in the framework of a close association, but may also have totally external causes, such as the appearance of a geographic barrier isolating two regions. The concept of co-speciation was initially developed to describe the joint speciation of hosts and parasites. The concepts of coevolution and co-speciation are used frequently in symbiosis research. Fundamentally, each organism promotes its own reproduction. As we saw in the first chapter, promoting the reproduction of another species to one's own detriment is not spontaneous [HAM 64]. Each partner in a symbiosis will therefore tend to try to move the cursor between costs and benefits to its own profit, even if it means damaging the other. This permanent conflict between cooperation and competition means that the association is never fixed. To stabilize the interaction, each partner must permanently adapt to the other. This idea that you must constantly move forward just to maintain the existing relationship, in other words to not move, bears the name of the Red Queen hypothesis in homage to the character from Lewis Carroll's "Alice in Wonderland", who says: "Now, here, you see, it takes all the running you can do, to keep in the same place. If you want to get somewhere else, you must run at least twice as fast as that!". Coevolution between host and symbionts is therefore the rule. The implementation mechanisms for recognition, dialogue and exchange between partners illustrate this. Since any change that arises in one partner can disturb or unbalance the relationship, the other is forced to adapt to it.

In reality, this equilibrium is never perfect and the partners are not equal. Most symbioses have asymmetries, to varying degrees. For example, coral, like most hosts of phototrophic symbionts, cannot survive without them, while most symbiotic algae, if they have retained their cell wall, are perfectly able to live outside the hosts [JOH 11]. This is also the case for chemosynthetic annelids such as *Riftia*, whose bacteria exist in a free-living state in the environment, whereas the hosts die if they lose their symbionts. Even symbioses that appear to be balanced, such as lichens, have asymmetries. The photobiont can live freely, sometimes even abundantly, while the mycobiont usually cannot [LUT 09]. As for intracellular symbioses with vertical transmission, like those of many insects and vesicomyid bivalves, they are obligate for both partners. This form of symmetry usually leads to strict co-speciation. However, they turn the bacterium into a mere factory, serving its host and under its absolute control. So is this a symmetrical relationship or a form of slavery? Can we even still talk of benefit for the symbiont, given that the absence of the free-living form is the direct consequence of the evolution of the association? Once more, we reach the limits of the classic definition of symbiosis, whence the need for a redefinition independent of the notions of cost and benefit. Asymmetry has significant consequences. For many facultative symbionts, the host is just one of many niches and natural selection applies over a whole life cycle, which may or may not pass through a symbiotic state. Free life also has the advantage of enabling them to meet other potential hosts. This has consequences for the evolution of symbionts and is reflected in many appearances and disappearances of symbiosis, as well as host shifts, in many symbiotic microorganism lineages.

Controlling symbionts enables hosts to adapt to external factors. There are six known or suspected defensive symbionts in the pea aphid [OLI 14]. Among them, it has been shown that in the absence of the parasitoid wasp against which it protects the host, the frequency of *Hamiltonella defensa* infection remains very low, at just a few percent of the aphid population. On the other hand, it fixes and infects almost all the individuals if the wasp is present [OLI 08]. In reality, it is observed in natural populations at intermediate frequencies. This time, symbiosis is facultative for the aphid but obligate for the bacterium. The evolutionary consequences for the host are nonetheless remarkable. Indeed, in insects, such facultative symbionts, known as "secondary", have a proven and important role in the speciation of their hosts. For example, by modifying the host's food preferences, they cause carrier individuals to settle on different plant species to those occupied by non-carrier insects [TSU 04]. In the long run, this contributes to the reproductive isolation of carrier populations and ultimately to the differentiation of new species [DAL 06].

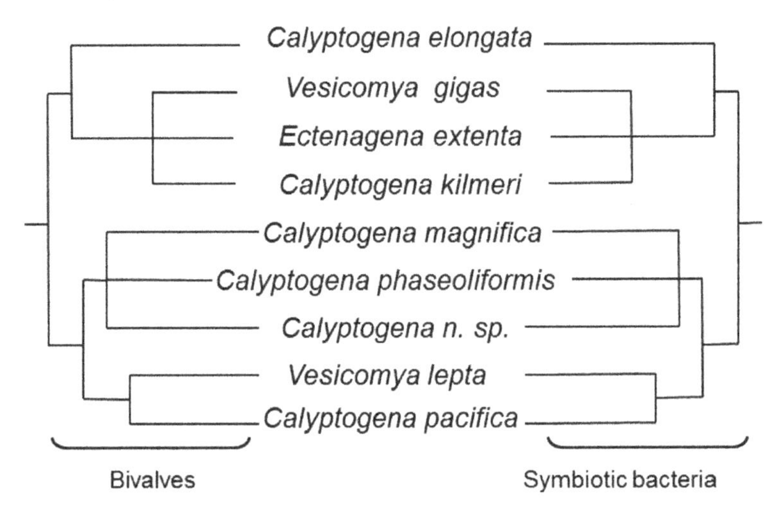

Figure 5.1. *Phylogenetic relationships between nine species of vesicomyid bivalves (left) and phylogenetic relationships between their associated symbiotic bacteria (right), adapted from Peek [PEE 98]. The dichotomies where one branch splits into two (e.g. V. lepta and C. pacifica) correspond to speciation events. Polytomies where one branch splits into more than two (e.g. C. magnifica, C. phaseoliforms and C. n. sp.) correspond to several speciation events so close in time that the analysis could not determine their exact sequence. In this example, the same topologies are found in the trees of bacteria and bivalves. A speciation event in one partner is therefore accompanied by a parallel event in the other: this is co-speciation. Co-speciation is widespread in symbioses with strict vertical transmission, as is the case for vesicomyid bivalves, but does not itself constitute proof of vertical transmission*

5.4. Morpho-anatomical consequences for hosts

The most recent definitions of symbiosis insist on the emergence of new functions. In many cases, these are accompanied by modifications of varying importance in the host's morphology, anatomy and physiology. Many organisms overdevelop pre-existing structures. This is the case, for example, for chemosynthetic bivalves such as *Bathymodiolus* and *Calyptogena*, in which the gill is hypertrophied until its surface area is more than 10 times larger than that of a non-symbiotic bivalve of equivalent size [DUP 16]. In xylophagous termites, a region of the digestive tube increases in volume to form the paunch that will contain the symbionts, just like the rumen in ruminants. These modifications are the product of coevolution, as shown by the evolution of herbivory in mammals [MAC 02]. Small ancestral forms were frugivores instead. Over evolutionary times, the growth trend in mammals is accompanied by a lengthening of the digestive tube, which increased the transit time of the alimentary bolus, enabling them to digest more refractory materials, such as grasses. At the same time, voluminous specialized digestion chambers, such as the rumen or cecum, which contain high densities of a symbiotic microflora that digests cellulose, developed. In ruminants, we observe an increase in the number of copies of the gene encoding lysozyme and its production, an enzyme normally involved in defense against pathogenic bacteria, but whose function here is to digest this microflora, which draws its carbon from plant matter. At the same time, whether it is a carnivore, an omnivore or a herbivore, endothermia (i.e. heat production and the maintenance of a high body temperature) also improved the efficiency of digestive symbionts by accelerating microbial metabolism. Indeed, it is observed that the growth optimum of a number of digestive bacteria is body temperature. Endothermia uses a lot of energy but improves digestive performance; overall, it is positive for the holobiont and has been retained by natural selection in the organisms concerned (mammals and birds).

In other organisms, new organs develop to harbor symbionts. The trophosome of the hydrothermal annelid *Riftia pachyptila* is formed from the animal's mesodermic germ layer as symbionts settle there and develop. It is composed of lobules, in which symbionts actively divide in the center, grow along the radius and are digested on the periphery (Figure 5.2). This tissue is very well irrigated, enabling symbionts to receive what they need, including reduced sulfur and oxygen, which are transported by a single, original hemoglobin capable of carrying both, thus preventing sulfur from being fixed instead of oxygen. We could also cite the bacteriomes of insects, groups of cells containing the primary endosymbionts, such as *Buchnera aphidicola* in aphids, or the crypts that contain the bioluminescent *Vibrio fischeri* in the squid *Euprymna scolopes*. In plants, we could cite nodules in legumes. Although the genetic determinants for trophosome formation in *Riftia* are still unknown, they have been identified, for example, in the aphid. The combination of developmental genes involved is unique to the bacteriomes, reinforcing the hypothesis of a true "new organ" [BRA 03].

In some cases, these structures develop only if symbiosis is established. It is obvious in legumes, where the nodule installation sequence has been well described and depends on the initial contact between partners. In lichens, the mycobiont takes different forms, depending on whether or not it is lichenized. It takes the globular form that makes a culture chamber for the photobiont only if the latter is present. In other cases, the symbiont-receiving structure pre-exists their installation. Aphids thus produce bacteriocytes in the bacteriomes, which are "prepared" before they are infected and express specific transcription factors for a new cellular lineage [MOY 08]. However, in individuals that have been artificially deprived of symbionts, we observe the secondary loss of these structures, which are therefore maintained only if infection occurs. Besides modified or completely new structures, other structures may decline or even disappear in symbiotic organisms, as in the case of the digestive tract in *Riftia*, which is present in the larvae but disappears in the adult annelid, which has no mouth or anus.

The genetic determinism of the installation or modification of structures is still little understood, since it requires both significant experimental work and the sequencing of large eukaryotic genomes with many billion bases. The sequencing of the genome of the aphid *Acyrthosphion pisum*, however, indicates that many of its genes are linked to symbiosis [SHI 11]. More generally, more and more studies are evoking a major role for symbiosis in various aspects of development, particularly in animals [MCF 13]. Symbionts provide signals that are useful at different developmental stages, including brain formation and epithelial renewal in mammals. The regulation of symbionts may be based in part on genes of bacterial origin present in the host genome and then expressed in the cells that contain them [MOR 14]. This is also the case in the aphid, whose genome contains genes of bacterial origin, which are abundantly expressed by the bacteriocytes containing *Buchnera*. These code, for example, for lysozymes, which are likely involved in the control of symbionts, as well as for the components of the cellular envelope, such as peptidoglycans, which are probably involved in the development of the envelope around the symbionts. However, paradoxically, these genes do not come from the genome of *Buchnera* itself, but from other bacteria, such as *Wolbachia* and *Rickettsia*.

We could again suggest an analogy here. We have seen that the eukaryotic cell's ability to integrate microorganisms is based on compartmentalization, making it possible to install the newcomer in an isolated subspace of the cytoplasm of the host cell. This makes it possible to communicate with it while avoiding disturbing the whole cell. In multicellular organisms, the presence of a symbiont is often linked to one or several tissues or organs that are specialized in this function. Once again, this is a form of compartmentalization, but at a higher level: that of an organ within a complex organism.

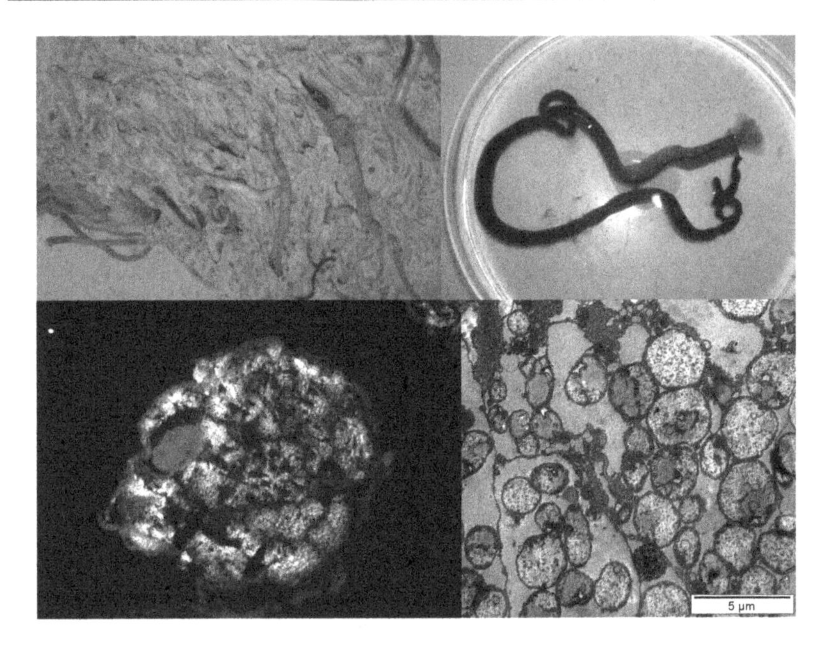

Figure 5.2. *Vestimentiferan annelids (here Lamellibrachia anaximandri from the eastern Mediterranean, related to Riftia pachyptila) live in chitinous tubes, from which only the pink gill protrudes (top left). The animal, which may grow to a meter in length, obtains oxygen and dihydrogen sulfide through its gill (top right). We see that it stretches out of its tube by means of a muscular part, the vestimentum (light brown), which gives its name to the group and enables the gill to be retracted inside the tube. The rest of the animal is a trunk, inside which the trophosome, the special lobular organ that contains symbionts, occupies the largest part (dark brown, the skin is transparent). At the end is a special region called the opisthosome, whose segmentation is similar to that of more classic annelids. The histological cross-section shows the trophosome filled with visible bacteria in white (bottom left, fluorescence microscope). These bacteria, which vary in size, are sulfur-oxidizing symbionts (bottom right, electron microscopy).*

5.5. Genetic and genomic consequences for symbionts

The consequences of symbiosis for the evolution of symbionts are particularly visible at the genetic level. They depend on the degree of dependence with respect to the holobiont, but also on possible exchange opportunities with other organisms. For example, in the human digestive tract, the rate of lateral gene transfer between bacteria reaches 25 times that observed in other environments. This phenomenon is explained by the diversity and high density of prokaryotes [MCF 13]. In this case, a direct consequence of symbiosis is the facilitation of the rapid emergence of new functions. An example of this is found in the human gut. There, a digestive bacterium was shown to have acquired the capacity to break down some typical algal polysaccharides by lateral gene transfer from a marine bacterium. This new capacity evolved within the Japanese population, which consumes a lot of algae. As we have seen, for facultative symbionts, the host is just one ecological niche among many. Natural selection applies to individuals in each niche. This contributes to the maintenance of functions that are indispensable in these various habitats. This is why *Rhizobium*, *Vibrio fischeri* and even the symbiont of the annelid *Riftia pachyptila* all possess the genes necessary for the production of a functional bacterial flagellum, which is expressed only during their free-living stage. Functions related to symbiosis are sometimes limited to a very small number of genes or located on the plasmids, just like the genes for nodulation and nitrogen fixation in some Rhizobia. In these conditions, symbionts are likely to evolve through both mutations and lateral gene transfer. For example, nodulation plasmids are regularly transferred laterally among soil bacteria.

As soon as the symbiont no longer has a free-living form independent of its host, gene loss phenomena are observed (Table 5.1). These losses are partly a consequence of the absence of a free-living stage, which renders some functions useless, such as the membrane protein or bacterial flagellum in *Buchnera aphidicola*. Genes that code for functions overlapping with those of the host, or even of another symbiont present, can also be lost. *B. aphidicola* has therefore lost the genes coding for the 10 non-essential amino acids that are produced by the aphid host. The inactivation of these genes through mutations, which transform them into pseudogenes, is no longer counter-selected. Their pure and simple loss, which leads to a reduction in genome size and therefore a decrease in the energy cost associated with its maintenance and replication, is even promoted, leading to genome erosion. Vertical transmission further accentuates this phenomenon, because only a small number of bacteria pass through the egg cells. This introduces a bottleneck into the life cycle of symbionts with each generation change of the host, increasing mutation rates (and therefore inactivation of useless genes, since the modification of important genes is counter-selected) and erosion rates, since bacteria with smaller genomes tend to divide more rapidly. The aphid and *Buchnera* share around 150–200 million years of common history, of which the very reduced genome of the symbiont as we know it today is the result (see Table 5.1).

Symbiont	Group	Host	Status	Trans.	Genome (Mb)	Genes
Escherichia coli	Gammaproteobacteria	-	L	-	4.6	~4,500
Rhizobium leguminosarum	Alphaproteobacteria	Leguminous plants	L and S, F	H	7.75	7,263
Vibrio (Aliivibrio) fischeri	Gammaproteobacteria	*Euprymna scolopes* (squid)	L and S, O	H	4.28	3,802
Endoriftia persephone	Gammaproteobacteria	*Riftia pachyptila* (vestimentiferan)	L and S, O	H	~3.7	?
Sulfur-oxidizing symbiont	Gammaproteobacteria	*Bathymodiolus septemdierum* (mussel)	L and S, O	H	1.47	1,471
Wolbachia pipiensis	Alphaproteobacteria	Insects	S, F	H and V	1.48	1,386
Wolbachia sp.	Alphaproteobacteria	Filiarial nematodes	S, O	V	1.08	806
Hamiltonella defensa	Gammaproteobacteria	*Bemisia tabaci* (silverleaf whitefly)	S, O	V	1.84	1,806
Vesiocomyosocius okutanii	Gammaproteobacteria	*Calyptogena okutanii* (clam)	S, O	V	1.02	939
Blattabacterium sp. BPLAN	Bacteroidetes	Cockroaches	S, O	V	0.641	581
Wigglesworthia glossinidia	Gammaproteobacteria	*Glossina* sp. (tsetse fly)	S, O	V	0.698	611

Buchnera aphidicola	Gammaproteobacteria	Aphids	S, O	V	0.64	564
Buchnera aphidicola (Cc)	Gammaproteobacteria	Aphids	S, O	V	0.416	357
Carsonella ruddi	Gammaproteobacteria	*Pachypsylla venusta* (hemipteran insect)	S, O	V	0.159	182
Tremblaya princeps	Betaproteobacteria	mealybug	S, O	V	0.139	121
Nasuia deltocephalinicola	Betaproteobacteria	*Macrosteles quadrilineatus* (cicada family)	S, O	V	0.112	137
Mitochondrion	-	*Cucurbita pepo* (gourd)	Organelle	V	0.982	28
Chloroplast	-	*Floydiella terrestris* (green alga)	Organelle	V	0.521	74

Table 5.1. *Characteristics of diverse bacterial symbionts and organelles: classification, host(s), status (L: a free-living stage of the bacterium exists; S: symbiotic bacterium; F: facultative symbiont for its host; O: obligate symbiont for its host), type of transmission (H: horizontal; V: vertical), size of genome in megabases (1 Mb = 10^6 bases) and number of genes. The bacterium Escherichia coli is indicated as an example that is representative of the classic size of the genome of a free bacterium. The last two lines are examples of mitochondria and chloroplasts with large genomes. These genomes are larger than those of many symbiotic bacteria, but their lower number of genes may be noted. It is observed that symbionts that can live in both free and symbiotic states generally retain larger genomes, in size and in number of genes, than obligate symbionts with vertical transmission. In the latter, genome reduction can be spectacular. Adapted from [MCC 12]*

In models of obligate endosymbiosis that have emerged more recently, the genome is less reduced and we have a better understanding of the mechanisms contributing to such an integration during evolution. In these more recent symbioses, we initially observe the abundance of sequences related to genome fluidity, which makes it possible to copy and paste sequence fragments from one place to another. This genome fluidity accelerates the modifications. However, these sequences related to fluidity tend to disappear during evolution and rapid genome erosion is observed [MOY 08]. The symbiont therefore loses functions and genomic fluidity. Genes related to the correction of errors during genome copying also tend to disappear [WER 02]. Mutations are therefore no longer reparable and, as they are most often detrimental, this transforms functional genes into non-functional pseudogenes, which will then be eliminated. Furthermore, it is observed that at this stage, chaperone proteins, such as GroEL, which prevent protein misfolding in order to make them functional, are greatly overexpressed, as if to compensate for the mutations. This again accentuates the loss of functions until a minimal genome is stabilized. At this stage, any supplementary mutation would be systematically counter-selected. If the symbiosis is highly integrated, the isolation of the symbiont inside the host cells limits its possibilities for exchanging genetic material with other microorganisms, and a stasis phase occurs. In *Buchnera*, over the last 50 million years, many base variations in the genome have been observed in the sequence, but none in the genome structure, i.e. how genes are arranged [TAM 02]. However, the story does not end there. In the extreme, the smallest *Buchnera aphidicola* genome (lineage Cc) contains only 416 kb and 357 genes, less than 10% the size of a typical bacterial genome. It has lost some of the genes necessary for the synthesis of amino acids essential to the aphid, but the missing amino acids are produced by a second symbiont that lives alongside *Buchnera* in the bacteriomes [PÉR 06]. The bacterial genome seems to be deteriorating and we may be directly witnessing this symbiont's replacement by the second, which complements it at the functional level. The replacement phenomenon may cause the extinction of the species of *Buchnera* living in this host. From this point of view, the highly integrated symbiosis may become an evolutionary dead end for this bacterium, which is so specialized that it no longer has any other possible ecological niche than its host. In symbionts for which the association is facultative, fewer modifications are observed than in those for which it is obligate, but symbiosis may also accelerate the evolution of the symbionts.

5.6. Integration between partners and the concept of organelle

If we wish to compare their characteristics, associations can be classified along a continuum of increasing integration, from separate organisms to a completely integrated entity, represented, for example, by the eukaryotic cell. The least integrated are facultative epibioses, in which symbionts colonize a surface of the host without the interaction going any further. The next step is epibioses controlled by the host, such as that between the flagellum *Mixotricha paradoxa* and the spirochetes that help it move, and then associations in which the symbionts live in the host's tissues but not in its cells, such as mycorrhizae. The development of a whole new organ by the host, such as the crypts in *Euprymna scolopes*, may represent a supplementary degree of integration, like the internalization of the symbionts in its cells. A more integrated degree of symbiosis would be the complete internalization of the symbiont, accompanied by its loss of autonomy in a dedicated organ, as is the case for *Buchnera*. Finally, the ultimate integration is the symbiont's transformation into an organelle, such as mitochondria or chloroplasts. This "gradation" is purely theoretical and does not in any case recreate the evolution of associations. In nature, symbioses appear, disappear and reappear, and the oldest are not necessarily the most integrated. Evolution does not necessarily tend towards increasing integration, which, as the previous section explained, may lead to the extinction of one of the partners. Furthermore, the concept of integration can be understood at many levels: genomic, cellular and metabolic [KEE 08]. However, such a classification has the advantage of categorizing the associations and identifying the various mechanisms at play in each, in order to reflect on the possible transitions between the various degrees. Furthermore, it highlights a sort of continuity between symbionts and organelles.

An organelle is a specialized cellular compartment associated with a membrane. Mitochondria and chloroplasts are considered to be two organelles of eukaryotic cells, rather than symbionts, despite their established bacterial origin. This transition from the status of symbiont to that of organelle is controversial. As a criterion, extreme genome reduction alone is not enough. Some symbionts, such as the bacterium *Carsonella ruddii*, have genomes of less than 160 kb, whereas those of some chloroplasts and mitochondria reach 500–1,000 kb (Table 5.1) [MCC 12]. A more relevant criterion is gene transfer from the symbiont to the nucleus of the host cell [MOY 08], accompanied by mechanisms that makes it possible to readdress the product of their expression back to the symbiont [KEE 08]. Most mitochondrion genes are therefore now found in the nucleus, just like many chloroplast genes. This phenomenon contributes to stabilizing and retaining the relationship. This definition seems more secure, but in the sea slug that practices kleptoplasty, mentioned in Chapter 2, alga genes whose products are directed to the plast have been identified in the nucleus of germinal cells lacking plasts, introducing the first example of photosynthetic gene transfer in a metazoan [JOH 11]. However, a recent genomic study of the host calls this gene transfer into question, observing that although they are expressed, the genes are not present in the genome of the egg cells [BHA 13]. More convincing is the example of aphids. The aphid genome contains nine intact genes received through lateral transfer from bacteria (although not from its symbiont *Buchnera*). At least one gene is expressed whose product is addressed exclusively to *Buchnera aphidicola* [SHI 11]. Another example is found in the insertion of almost the entire *Wolbachia* genome into chromosome 2 of its host fly *Drosophila ananassae* [WER 08]. Thus, even gene transfer is not a watertight criterion for distinguishing an organelle from a symbiont, and the existence of a line between the two is increasingly called into question [MCC 14]. Just as for symbiosis in general, no definition is really able to distinguish between associations, and it is important to be aware of this.

5.7. The origins of the eukaryotic cell

Beyond the question of organelles is that of the origin of the eukaryotic cell and the transition between the prokaryotic and eukaryotic states. The eukaryotic cell is characterized by its nucleus and by the mitochondria present in most current eukaryotes. It is now recognized that mitochondria derive from an endosymbiotic alphaproteobacterium present in the common ancestor of all eukaryotes, although the initial nature of the interaction (oxygen detoxification, syntrophy) is, as we have seen, a matter of some debate [BAU 14]. It is also recognized that the nucleus was formed in a prokaryotic cell through membrane invagination, either in the ancestral cell during phagocytosis of another prokaryote when the compartment first developed, or through a form of cell fusion. The nature of this initial host is debated, and the candidates may be Archaea or various types of bacteria. The organism that supplied the nucleus is also a matter of debate: archaeon, spirochete bacterium or even virus. There are several hypotheses as to the initial success of this association. One hypothesis proposes syntrophy between partners exchanging substrates, a phenomenon of which we see many current examples in nature, which leads to highly connected consortia. An example is provided in the next chapter [MOR 98]. The nuclear genome of eukaryotes itself probably represents the combination of a proteobacterium genome and an archaeon genome. The former may have contributed most metabolism genes, while the second contributed genes encoding structural proteins associated with DNA, replication, transcription and translation systems.

In the classic model defended by Lynn Margulis at the beginning of the 1970s, these various compartments are the product of the internalization within a host prokaryote of external partners through an endocytosis mechanism (see Figure 2.1, primary endosymbiosis). In this model, known as an "outside-in" model, the capacity for endocytosis pre-exists the original eukaryotic cell. In 2014, David and Buzz Baum proposed an original alternative model. Observing the propensity of some Archaea to produce protrusions towards the exterior of the cells, also known as cytoplasmic extensions, they suggested that these protrusions might be able to stretch out to completely surround the mitochondria-like prokaryotes located on the surface of the original cell and then merge together to isolate these symbionts from the exterior (Figure 5.3). They are thus incorporated into the cellular space. This time, it is not the symbionts that enter the host (the "outside-in" model) but extensions towards the exterior of the host cell that surround the symbionts, also known as an "inside-out" model. They provide many arguments in support of their theory. For example, a model in which membranes form protrusions would explain the complex organization of the intracellular membranes observed in the endoplasmic reticulum of the Golgi apparatus. The "inside-out" model leads to several interesting hypotheses, such as the idea that the initial host may have been an archaeon (capable of producing these extensions) and that natural selection may have promoted the engulfing of the symbionts because it increases the exchange surface between them. Although the prokaryotic origin of various compartments of the eukaryotic cell has been demonstrated beyond reasonable doubt, it will perhaps never be possible to decide between the "outside-in" and "inside-out" hypotheses. Nonetheless, this example illustrates the dynamism and relevance of research carried out on the subject and produces new interpretations of cell biology. Furthermore, it is clear that the eukaryotic cell itself is a holobiont, a composite organism combining the genomes of several very different partners.

Figure 5.3. *The "inside-out" model according to David and Buzz Baum. The symbionts live on the surface of the future host cell ("eocyte"). This cell emits protrusions (left) that gradually surround the symbionts (center), increasing the exchange surface between the partners. Eventually, these protrusions will totally surround the symbionts and fuse to become a cytoplasm around a protonucleus, producing a complex set of membrane folds, as we see in current eukaryotic cells (right)*

In the Archaeplastida, the eukaryotic cell will gain chloroplasts, which, as we saw in Chapter 2, derive from a cyanobacterium present in the ancestor of this lineage. But this happens much later, in a cell that is already a eukaryote. We may then wonder whether the appearance of new organelles is still possible and if this might be the final evolutionary trajectory for some of the symbioses described here. For many experts, there are several conflicting points, at least in multicellular organisms. First, symbionts are usually located in specific compartments within the organism (organs, tissues), rather than in all the cells. This suggests that if massive gene transfers occurred from a symbiont to the nucleus of the multicellular host, these genes would immediately need to be placed under the control of transcription factors, which would limit their expression to these organs at the risk of disrupting the entire system. This seems fairly unlikely. Furthermore, the complexity of the developmental mechanisms introduces significant risks of interference. Therefore, in principle, the answer would be no, at least for organelles as we know them (mitochondria, chloroplasts). However, we should bear in mind the example of the aphid, which expresses some genes of bacterial origin, now located in their own genome, in symbiotic tissues. This reminds us that genetic integration pathways are likely to be still open and offers proof that the line between symbiont and organelle is (and will likely remain) fluid.

5.8. Symbiosis imposes a new perspective on the tree of life and on evolution

Through the acquisition of new capacities, symbiosis is a factor in adaptation and a driving force behind evolution. The evolution of species is commonly represented as a phylogenetic tree, depicting the relationships between organisms as branches that divide over time, as shown in Figure 5.1. The paradigm is the dichotomous tree. One branch represents one lineage over time. It may interrupt if the lineage becomes extinct. It can also maintain, or split into two or more branches during speciation events. Phylogenies can be inferred based on many types of characters. The most commonly used today are the alignments of DNA sequences, whose length varies from a few hundred bases to entire genomes. We therefore estimate evolutionary relationships between organisms by comparing the similarity of their genes. But the gene phylogeny is not necessarily that of the species that harbor them. Some genes are poor markers and cannot resolve relationships because they vary too much or too little over time. Others are not reliable because their destiny is separated from that of the species, as is the case for genes that are passed from one species to another unrelated species through lateral transfer. In this case, a branch in the tree links the donor species and the recipient species, which does not reflect the existence of a common ancestor.

Symbiosis can also be problematic to integrate into the tree example. To describe the evolution of holobionts formed from several organisms using a tree, its branches must be able to divide, as is the case in the classic model, but also to rejoin and fuse to bring about the creation of a new holobiont assembled from two lineages, the partners (Figure 5.4). The question of whether eukaroytes are closer to Archaea or bacteria illustrates this. The eukaryotic cell is likely the result of the fusion of an archaeon and a bacterium. Classic phylogeny methods will thus bring together the eukaryote lineage and that of the two prokaryotic lineages that has contributed the most to the genome of the common ancestor of all eukaryotes. We might wonder whether such an interpretation makes sense, given that the true founding event is the fusion itself. Such phenomena are only just beginning to be taken into account under the name of reticulate evolution.

Finally, we can consider the very notion of species. In 1992, in his article describing the concept of a holobiont, Mindell proposed that each partner retains its species status but that the holobiont itself, as a single entity, constitutes a species [MIN 92]. In this proposal, species behave like Russian dolls, some incorporated inside others. Although this idea may seem provocative, it fits well with our knowledge of the origins of the eukaryotic cell and is not new in practice, since for more than a century, each lichen, for example, has received a genus and species name, despite being a holobiont rather than a unique organism. This name is, however, nowadays attributed only to the mycobiont.

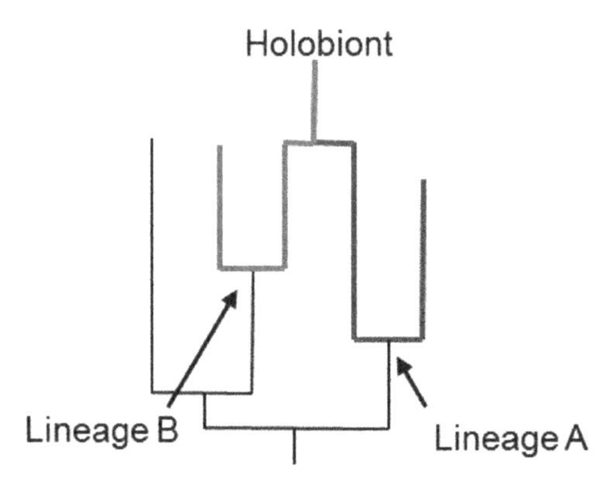

Figure 5.4. *In principle, a phylogenetic tree includes branches that split from one lineage into two, each reflecting one speciation event. The assemblage of a holobiont through the association of lineage B (red) and lineage A (blue) is represented here by the meeting of two branches, this time reflecting the fusion of two lineages (pink). Adapted from [MIN 92]. For a color version of this figure, see www.iste.co.uk/duperron/symbioses.zip*

5.9. Are the concepts of holobiont and hologenome ultimately useful? The story continues...

The very concepts of holobiont and hologenome are now being called into question by some authors. The two concepts enable us to make a certain number of predictions, and a concept of evolution based on the hologenome has been developed from the hypothesis that the holobiont is the most relevant unit of selection [BOR 15]. They have gained in popularity over the last few years and are increasingly influential to the way researchers study symbioses, hence their frequent use (including in this book). However, the concepts themselves have been the subject of recent controversies. Although at first glance the holobiont seems relevant as a unit on which natural selection acts, some authors insist that hosts and symbionts do not, in fact, necessarily evolve as a single unit. Indeed, a recent article by Douglas and Werren observes that if fidelity between partners is low, as is the case in many facultative and horizontally acquired associations, the hologenome content changes from one generation to the next and the holobiont is not a unit of selection [DOU 16]. Furthermore, there are conflicts of interest between partners. A microorganism living with or without a host is subject to selection in both cases and its fitness will depend on its performance in these two habitats (see Figure 1.3). Its evolution will therefore not be governed only by its interaction with a host. In vertically transmitted symbioses, the example of *Wolbachia*, which manipulates the reproduction of its host insects to promote females (the only ones able to transmit *Wolbachia* to their offspring), also illustrates this conflict, in which natural selection promotes the bacterial lineage most suited to this manipulation, and therefore not necessarily the most favorable to the host's fitness. Within the holobiont, there may also be conflicts between different symbionts themselves, which can lead to replacements. Douglas and Werren believe that there are units of selection inside the holobiont that must be taken into account, and that it cannot therefore be treated as a single unit of selection. For these authors, the idea that the holobiont (and its hologenome) evolves as a unit is false, because it does not include this conflictual aspect of the relationship between partners or take account of the fact that, in reality, selection acts on different levels, even within the holobiont. They believe these terms are useless and should be abandoned or replaced with more neutral ones, such as "symbiome" and "symgenome", which acknowledge the association without assuming that it has evolved as a unit [DOU 16]. Without taking sides or jumping to conclusions about the future of the controversy, this ongoing discussion illustrates the liveliness of the debates that are currently gripping the scientific community.

The diversity of the symbioses, of the players involved, of the variety of their interaction and evolutionary origins indicates unequivocally that symbiotic association between organisms is more the rule than the exception. Through a common language, the association is likely easy to establish, as it often requires only a few changes and most lineages have the necessary tools. The similarities between the mechanisms used by beneficial symbionts and pathogens show that, in reality, there is a continuum of relationships. The evolutionary paths of the various associations differ and most association attempts have probably not been retained by natural selection. But those that we observe today are proof of their success. Evolution through mutation and natural selection is gradual and proceeds with small steps, meaning that the acquisition of a new competence can take millions of years. However, evolution through the association of organisms proceeds in leaps. Over short periods of time, it can produce completely new and relevant functions likely to open up new ecological niches and thus contribute to the diversification of life forms. It can be considered similar to evolution through lateral gene transfer, which enables prokaryotes, bacteria and Archaea to acquire whole metabolic pathways from non-related species. We can take the analogy even further, suggesting that the rapid acquisition of new functions occurs predominantly through lateral gene transfer in prokaryotes and through association with other organisms in eukaryotes. This difference is based on the distinct abilities of these two groups. Prokaryotes are very good at absorbing, integrating and expressing foreign DNA, while eukaryotes are particularly effective at absorbing and integrating whole cells. Events such as the origin of the eukaryotic cell demonstrate to what extent symbiosis is a significant driving force behind evolution.

The significance of symbiosis in evolution has led Douglas Zook to call it "evolution's co-author" in the title of a recent article [ZOO 15]. Perhaps it is fairer to say that it is one important driver among several; nonetheless, this provocative title has the merit of bringing the concept of symbiosis back into the heart of biology and evolutionary sciences, where it has long been treated as a phenomenon of minor, almost anecdotal, importance.

6

Symbiosis and the Biosphere

6.1. Symbiosis and the current biosphere

6.1.1. *Symbiosis and ecosystem productivity*

Among their most important roles, symbiotic associations are involved in nutritional functions and therefore contribute directly to various biogeochemical cycles. For carbon, for example, all photosynthetic eukaryotes have chloroplasts of symbiotic origin. On land masses, almost all primary production is carried out by vascular plants. This represents around $4{,}700 \times 10^{12}$ moles of carbon fixed per year in total, and primary production can exceed 2 kg per square meter per year in the most productive tropical forests. In the oceans, the productivity per unit area of large algae is comparable, but it is phytoplankton that are primarily responsible for the equivalent total primary production, with around $4{,}040 \times 10^{12}$ moles of carbon fixed per year in total [CAN 05]. Here, eukaryotes are once more dominant in number. Terrestrial or aquatic, they are responsible for most of the 285 billion tonnes of biomass produced annually and therefore for most of the carbon input into food chains. Even at the bottom of the oceans, chemosynthetic primary production is mainly performed by bacteria living in symbiosis with the invertebrates that colonize cold seeps and hydrothermal vents, although no estimate of their overall productivity has yet been produced.

Plants associated with nitrogen-fixing bacteria are responsible for the conversion of atmospheric nitrogen into molecules that can be used by their hosts and then by herbivores, and therefore for its entry into food chains accessible to terrestrial organisms. Mycorrhizae contribute to the mobilization of phosphorus stores that are difficult to access, to the advantage of plants and to the food chain as a whole, and can make up 80% of the microbial mass of the soil, also helping to stabilize its structure.

Coral reefs, thanks to the interaction between corals and zooxanthellae, and deep-sea hydrothermal vents, with their invertebrates associated with chemoautotrophic bacteria, are two of the most productive marine ecosystems and are both sustained by symbiosis. On land, the combination of carbon and nitrogen fixation and phosphorus acquisition through symbiotic associations is behind the extraordinary productivity of forest ecosystems. When we look at them more closely, therefore, many of the most productive ecosystems on our planet per square meter are based on symbioses.

6.1.2. *Symbiosis and biodiversity*

Since the Convention on Biological Diversity was signed at the 1992 Earth Summit in Rio de Janeiro, biodiversity and its preservation have become major public policy challenges. The concept of biodiversity covers the diversity of species and their genetic traits within a given area. It therefore perfectly complements the notion of ecosystem productivity. Symbiotic interactions contribute greatly to this biodiversity, directly or indirectly. Indeed, many of the symbiosis-based ecosystems mentioned previously are also biodiversity hotspots. According to the *Census of Marine Life* (coml.org), coral reefs harbor between a quarter and a third of marine species across only 2% of the surface of the oceans [PLA 11]. Tropical rainforests, which cover around 7% of the surface of our planet (around 13% of land), contain around half the species in the world. Although these figures are naturally approximate and estimates vary according to the sources, it is clear that these ecosystems are particularly favorable to the coexistence of many species. Symbioses therefore play an indirect role. They are responsible for the productivity of these ecosystems, but it is the diversity and heterogeneity of the habitats, which can be seen as a favorable secondary effect, that provides an endless supply of available ecological niches.

Symbioses have a more direct effect on the biodiversity of other ecosystems. Their highly diversified roles help holobionts to adapt to many ecosystems from which they would otherwise be absent and often contribute to their diversification. Lichens, for example, are often the first multicellular organisms to establish themselves on newly formed or "sterile" landmasses, as has been observed following volcanic eruptions. They promote the formation of a soil that will then receive other species. Some evolutionary radiations are also a direct consequence of symbiotic innovations, which cause rapid diversification in many taxa. For example, nutritional symbioses have enabled various insect groups to adapt to very unbalanced and specialized diets, such as the consumption of elaborated sap or wood, as discussed in Chapter 2. We have also seen the role of some secondary endosymbionts in insects in the choice of the plant that their host will inhabit. This choice may lead to the reproductive isolation of these insects from their counterparts without the same secondary symbiont, and ultimately to the appearance of new species. We might also cite the example of mussels living in chemosynthesis-based deep-sea environments. Here, it is the appearance of symbiosis with sulfur-oxidizing bacteria that has enabled these habitats to be colonized, on which both hosts and symbionts rapidly diversified.

Finally, the symbiotic lifestyle is itself a source of biodiversity. It is estimated that 90% of the bacteria in the digestive tract of xylophagous termites, as well as the majority of the flagellate species, exist only there. Among the bacteria, one group is so different from all the others that it has even been established as a phylum, the Endomicrobia [STI 05]. Another bacterial phylum, the Poribacteria, exists only in association with sponges [KAM 14]. In fact, a host is always a very specific habitat, to which symbionts must adapt and within which they can sometimes diversify to a remarkable extent. In vertebrates, for example, a general trait seems to be possession of very complex microbial consortia [MCF 15]. The mere presence of vertebrates in an ecosystem is therefore necessarily accompanied by a significant diversity in the associated microorganisms, whatever their roles.

6.2. Symbiosis and history of the biosphere

6.2.1. *Syntrophy and symbiosis*

Life on Earth probably appeared over 3.5 billion years ago and the metabolisms of the first life forms diversified rapidly. It is, however, difficult to date the origins of symbiosis phenomena. However, strong interactions between prokaryotes, associated with close contact between them, probably existed at a very early stage, notably in their nutrition. Indeed, in current nature, we know of many associations in which prokaryotes with complementary metabolisms associate and provide each other with the substrates they need. This phenomenon is called syntrophy and can be described as an obligate mutualist metabolism [MOR 13]. Mutualism is based on the provision of one chemical compound by one partner, consumed by the other in exchange for a reward. The anaerobic oxidation of methane by microbial consortia is a good example of this phenomenon [KNI 09]. Many mechanisms are known today, all involving methanogenic Archaea [JOY 12]. In principle, these Archaea produce methane in anaerobic conditions. However, in anaerobic methane oxidation, methanogenesis works in the opposite way, consuming methane and producing carbon dioxide. Most of the mechanisms identified are based on an obligate syntrophic relationship between these Archaea and bacteria. In the first case identified (and the best documented to date), these bacteria are sulfate-reducers: they reduce sulfate to sulfur by transferring electrons to it directly. These electrons are provided by Archaea and come from methane oxidation (Figure 6.1). The electron transfer gradient makes methanogenesis work in the opposite way, oxidizing methane into carbon dioxide in the absence of oxygen, a reaction that would not occur spontaneously. In another case that has been identified, the bacterial partners are different and anaerobic methane oxidation is this time coupled with nitrate reduction. The energy yield is modest, but this is offset by the abundance of methane available in certain habitats, such as cold seeps in the deep ocean, where chemosynthetic animals also live. Here, the obligate aspect is related to the fact that syntrophic activity leads to anaerobic methane oxidation, which would be not be favored if the partners worked separately. Mutualism is necessary for the creation of this metabolism, but not necessarily to the survival of the partners. In general, partners can live without one another by relying on other types of metabolism. However, the association enables the consortium to exploit a new ecological niche, in this case methane-rich anoxic sediments.

Far from being anecdotal, syntrophy phenomena can have a strong impact on the biosphere. For example, anaerobic methane oxidation in marine sediments is today responsible for the consumption of 90% of the methane produced in the oceans, making up between 7 and 25% of its worldwide production. It is therefore reasonable to believe that such consortia played a significant role in the formation of the biosphere at a very early stage. We may wonder about the relationship between syntrophy and symbiosis. As it does not produce new structures and usually seems to be facultative for the partners, syntrophy does not lie within the framework of the classic definitions of symbiosis, but it does for many authors who use broader definitions [MOR 13]. The question has not been resolved, but either way, syntrophy relationships show us what the first associations between prokaryotes may have looked like.

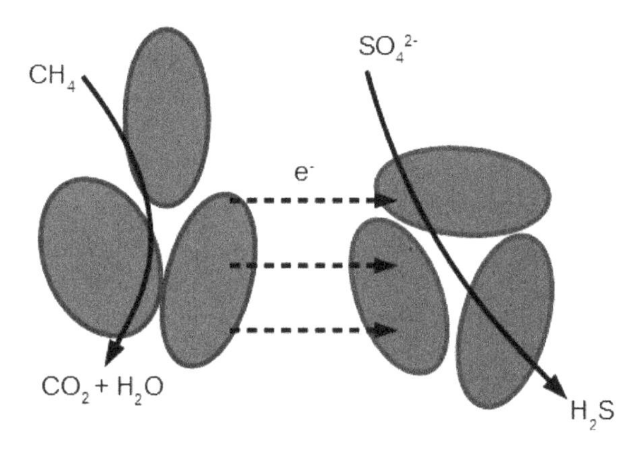

Figure 6.1. *Anaerobic methane oxidation mechanism. The archaeon from the ANME group (left) oxidizes the CH_4 methane in the absence of oxygen. The bacterium from the deltaproteobacteria group (right) reduces the SO_4^{2-} sulfate to H_2S sulfur. The two reactions are a pair, the electrons needed by the bacterium coming directly from methane oxidation by the archaeon. This electron transfer between physically associated partners makes methanogenesis work in the opposite way, leading to methane oxidation in anaerobic conditions. The interplay between partners creates a metabolism that is otherwise not favored: this is syntrophy. Diagram from [JOY 12]*

6.2.2. *Origin and diversification of eukaryotic lineages*

We saw in the previous chapter that the eukaryotic cell originated in a symbiosis between prokaryotes, perhaps also explained by syntrophic interactions [MOR 98]. The appearance of the eukaryotic cell between 1.6 and 2 billion years ago was a major evolutionary transition in the history of the biosphere. The diversification of lineages had a considerable impact. The size of organisms increased, and the common ancestor of all current eukaryotes was already a large cell. Besides the nucleus and the mitochondrion, whose origins have already been discussed, the cytoskeleton is an important characteristic of the eukaryotic cell. Once seen as the result of the association with a bacterium from the spirochetes groups, its origins remain unclear. It is, however, the cytoskeleton that gives the eukaryote its shape and mobility and enables intracellular transport, such as endocytosis [KAT 12]. The genome of the first eukaryotic cell was chimeric, in other words made up of genes that came from "founding members" of this consortium, including at least one bacterium and one archaeon. Quickly, unicellular eukaryotes, still called protists, diversified, first in the form of heterotrophs, their size reaching more than one hundred times that of the prokaryotes on which they likely fed. They could be presented as the giant super-predators of this period. Nonetheless, their ecological significance before the acquisition of the capacity for photosynthesis must have been relatively limited, although the fossil record is poor for this period. It was considerably enriched with the appearance of often photosynthetic forms producing biomineralization, such as radiolarians, foraminiferans and coccolithophorids.

6.2.3. *Phototrophic symbioses and atmospheric oxygen*

The initial acquisition of the chloroplast over a billion years ago by a eukaryotic ancestor of the Archaeplastida was followed by many secondary and tertiary endosymbioses in which a photosynthetic eukaryote was itself integrated into a host (see Figure 2.1). Acquisition of photoautotrophy by eukaryotes is the source of the immense diversity of the current photosynthetic lineages, ranging from minuscule unicellular algae to large forms, such as kelp and giant sequoias. This symbiosis is linked to the history of oxygen on our planet [CAN 05]. Until 1.9 billion years ago, the oxygen produced by phototrophic cyanobacteria since their appearance, at least 2.7 billion years ago, was totally absorbed by various reactions and thus probably rare in the atmosphere and in the oceans. After this period, oxygen, then toxic to many anaerobic organisms, began to accumulate in the atmosphere, until it reached around 10% of its current value. It was only 850 million years ago, i.e. after the appearance of chloroplasts in eukaryotic cells, that oxygen levels in the atmosphere rapidly increased. They peaked 300 million years ago, making up 35% of the air, due to the increase in primary production by photosynthetic eukaryotes, first in the oceans and then on land. As well as their role in the oxygenation of the atmosphere, unicellular phototrophic eukaryotes that produce mineralized elements are also responsible for vast sedimentary deposits of carbonate rocks, such as coccolithophorid tests, which, for example, make up the majority of large chalk deposits in the Paris Basin.

6.2.4. *Symbiosis and the origins of pluricellularity in animals?*

Animals appeared between 700 and 800 million years ago. Their closest relatives are choanoflagellates, unicellular eukaryotes that move using a propelling flagellum and trap particles and microorganisms using small cellular extensions that form a collar around the flagellum. Very similar cells are found in sponges, which use this type of cell (called a choanocyte) to create currents and filter water. In the choanoflagellate *Salpingoeca rosetta*, the formation of colonies through cellular division, which can be considered a transition to the pluricellular state found in animals, has been observed [FAI 11]. It is clear that some bacteria induce the formation of these colonies, but the exact mechanism behind this is not yet understood. On this subject, it is interesting to observe that sponges, the first animal group to emerge, are very rich in various bacteria, which can make up 50% of their mass [WEB 12]. We might therefore wonder about the possible role of the association with bacteria in the establishment of pluricellular structures, and even of pluricellularity itself, in the course of metazoan (i.e. animal) evolution. This is no more than a hypothesis at this stage, but it is significant, since it highlights the importance of including interspecific interactions in any study of evolution.

6.2.5. *Contribution to the appearance of terrestrial plants*

The oldest fossils of land plants (or embryophytes) appeared around 470 million years ago, although some molecular dating studies suggest that they are even older. Besides the gradual evolution of new structures that enable them to exploit the ground–air interface while retaining water (roots, leaves, vessels carrying sap, wood, etc.), the colonization of the continental habitat by terrestrial plants seems to owe a lot to the mycorrhizae that enable them to use soil nutrients. Indeed, although fossilization of fungus hyphae is not very common and fossils are difficult to interpret, unambiguous traces of arbuscular mycorrhizae are abundantly present in association with fossil plants found near Rhynie in Scotland and dated to 407 million years ago. These mycorrhizae are associated with structures that are not yet true roots [SEL 15]. Molecular dating converges, suggesting that the common ancestor of terrestrial plants may have already had mycorrhizae. At another level, the resistance to drought conferred by endophytic fungi could also have facilitated the colonization of terrestrial habitats [ROD 08]. The observation that all endophytes improve the host plants' water usage is a sign that this may be an original characteristic of these fungi. The productivity of terrestrial plants, previously limited by nitrogen availability, later benefited from nitrogen-fixing symbionts, the first of which is believed to have appeared around the end of the Cretaceous period, 65 million years ago. The colonization of land by lichens is less well documented, since fossils are rare and sometimes controversial. The oldest fossils on which there is consensus date from only 415 million years ago, leaving some doubt about the primacy of terrestrial plants or lichens in the colonization of continental masses.

The appearance of terrestrial plants able to colonize continents accelerated the accumulation of oxygen in the atmosphere and contributed to the aforementioned peak in atmospheric oxygen registered in the carboniferous period.

We could cite many more examples. The phenomenon of symbiosis has shaped many aspects of the biosphere as we know it, since it is the source of innovations related to as many leaps in the course of our planet's history. The most significant is probably the appearance of the eukaryotic cell, but other symbioses are also behind the exceptional productivity of many ecosystems. Today, these ecosystems are seen as carbon sinks that could buffer the rising atmospheric concentrations linked to anthropogenic emissions. Symbioses are also a significant factor in the diversification of species. On the other hand, the extinction of a single animal or plant species can lead to the "invisible" extinction of a large number of its symbiotic species for which this organism is the sole possible habitat. Our planet could be called a symbiosphere, to highlight the importance of this phenomenon. In our century, in which environmental changes caused by human activities are accelerating, our ability to anticipate the consequences of these changes relies on complex models that should take more account of the nature, workings and fragility of symbiotic interactions in order to refine their predictions.

7

Good Uses for Symbiosis

A century and a half of research into symbioses has brought its share of discoveries and re-evaluations. New associations, new types of partners and new roles are outlined each year. This is the daily bread, as it were, of the discipline. However, the contribution of such studies to our understanding of living organisms is not limited to expanding a catalogue of symbiotic relationships. Recent discoveries are opening up new avenues for research that are only just beginning to be explored, in both the fundamental and applied fields, in very varied ways. Here are a few examples.

7.1. Some avenues in fundamental research

7.1.1. *Viruses: new players in symbiosis*

To begin with, different authors disagree as to whether or not viruses are living organisms [MOR 09a]. This harks back to the very definition of living organisms and to the highly specific properties of viruses, which hijack the machinery of the cell that harbors them in order to multiply. However, this debate lies beyond the scope of this book. In Chapter 3, we saw the example of *Hamiltonella defensa*, a facultative symbiont of the aphid that protects it from the parasitoid wasp *Aphidius ervi* by producing toxins that are fatal to its larvae. The genes encoding these toxins are located on a bacteriophage [CLA 14]. In this case, the integration of the virus genome into that of the bacterium is responsible for its defensive function. Another well-documented example is the infection of mice by gammaherpesvirus 68 in its latent form. This virus protects the mice against many pathogenic bacteria, including the plague agent *Yersinia pestis*. Although the virus remains potentially detrimental itself [BAR 07], protection is a secondary effect here. Indeed, the presence of the virus stimulates the host's immune system and, in particular, the production of interferon and the activation of macrophages, which limit the effect of the bacterial pathogen. Another example that has already been mentioned is the heat tolerance that some endophytic fungi confer on their host plant. In reality, this can be attributed to a virus present in the endophyte and the tolerance disappears if the virus is eliminated [MÀR 07]. In placental mammals, the development of a syncytium, i.e. a group of cells that are not separated by membranes and therefore share a single giant cytoplasm, in the placenta requires specific proteins that originate in retroviruses integrated in the genome of animals [ROO 11]. These examples, along with some others, show that the study of the role of viruses as potential symbionts and not only as pathogens is certainly an avenue for future research.

7.1.2. *Symbionts as models for synthetic biology*

The aim of synthetic biology is to apply engineering methods to biotechnology, in order to shape living organisms so that they meet precise specifications. This may involve producing an artificial cell. Although this aim may seem ambitious, perhaps even arrogant, a group led by Daniel Gibson and Craig Venter (also known for contributing to the sequencing of the human genome and of many model organisms in biology) managed in 2010 to synthesize a whole bacterial chromosome and make it take control of a cell whose own genome they had first removed [GIB 10]. This “grafted” synthetic cell was able to reproduce and had the expected phenotype given its genome. In fact, the authors synthesized *de novo* a genome that was absolutely identical to that of *Mycoplasma mycoides*, a pathogenic bacterium of ruminants whose genome contains 1.08 million bases. This classifies it among the bacteria with the smallest genomes, but it is still much larger than the genomes of many obligate endosymbionts, such as *Buchnera aphidicola*. This is what makes some symbionts interesting models. In the quest for the minimum genome required to obtain a living organism, symbionts with micro-genomes are particularly interesting, such as *Carsonella ruddii* and its 159 kb genome, *Nasuia deltocephalinicola* with 112 kb and *Tremblaya princeps* with 139 kb. The latter also has the distinctive feature of itself containing a symbiotic bacterium (*Moranella endobia*), making it the only known example of symbiosis of one bacterium within another [MCC 12]. This minimum genome, which would, of course, function only in nutrient-rich and stress-free conditions, is the holy grail of synthetic biology, because it would provide a list of the only genes that are essential to life. On this basis, we could build simple organisms through bioengineering, minimizing the risk of interference with the specific functions that we wish them to perform.

There is at least one other field of synthetic biology in which symbiosis is a source of inspiration: modularity [POR 13]. Many systems, from computer programs to industrial processes, are broken down into modules that are separate and relatively independent from one another. This design enables them to be modified or improved in isolation, without the risk of causing a chain of disturbances in another part of the program or system. On the other hand, a module may potentially be transferred as it is to another system. For example, synthetic biology uses sequences as building blocks:

promoters, functional genes and terminators. New properties are born from combining them. In living organisms, such combinations are sometimes present on plasmids, modules whose acquisition by an organism leads to that of the associated functions. Symbioses also work a bit like this: each partner has its own evolutionary history and the acquisition of a symbiont amounts to the acquisition of a new biological module, a new block. In the case of strict endosymbioses, such as that between *Buchnera aphidicola* and the pea aphid, the loss of many genes in the course of evolution leaves space for functional complementation phenomena, which may be seen as the optimization of the integration of the module, but compartmentalization continues. Symbioses provide an example of how we might envisage constructing new functions through a modular approach and inspire various projects that aim to develop microbial consortia through synthesis or from modified symbionts.

7.2. Some avenues for applied research

Symbiotic organisms are already widely used in biotechnologies, notably for the production of interesting molecules. For example, many compounds are extracted from lichens, including odorant molecules, for which the perfume industry uses 8,000 tonnes of *Pseudovernia furfuracea* and *Evernia prunastri* lichens annually [LUT 09]. Other molecules are studied for their potential in human health. Bryostatins are produced by the bacterium *Endobugula sertula*, an endosymbiont of the bryozoan *Bugula neritina*, and play a role in repelling predators. They are also powerful antitumors that may be useful in the fight against cancer, and they improve memory in rats, also making them candidates for the fight against Alzheimer's disease. Clinical trials are ongoing. Alongside these already identified uses, current research into symbioses is rich in promise for the future in a wide range of fields.

7.2.1. *Unexpected roles of the human microbiota*

The human-associated microbiota has many roles. The most studied is the digestive tract microbiota, the complexity of which was discussed in Chapter 2. Its main role is to help digest food and, in particular, to produce vitamins. Nonetheless, over the last few years, it has been discovered that the role of microbial communities does not end here. Through their regular presence in our digestive tract, microorganisms also restrict the space remaining for those that cause diseases, both by occupying the space and by producing compounds that are anti-inflammatory or toxic to others [MAZ 08]. We have also discovered the role of this microbiota in the formation and "education" of the immune system when it comes to recognizing intruders. Many recent studies comparing the microorganisms associated with individuals suffering from various pathologies to those associated with healthy subjects have clearly demonstrated the link between some diseases and irregularities in digestive flora. This is the case for pathologies such as obesity and Crohn's disease. In the case of obesity, it has been shown that the microbiota of obese mice and humans are characterized by a lower diversity of bacteria and more efficient use of refractory compounds in digestion. These communities take more energy from the alimentary bolus than they do in slim individuals, leading to an increase in fat mass [TUR 09]. The link between microbiota composition and weight was demonstrated when researchers were able to make mice lose weight simply by transferring the intestinal microbiota of slim humans to them [RID 13].

More surprisingly, this seems to also apply to much more unexpected pathologies, such as certain depressions, in which the link between digestive tract bacteria and nervous system bacteria is only just beginning to be explored [END 15]. This is related to the presence of large quantities of nerve cells around the intestine, sometimes called the "second brain", in direct contact with substances of microbial origin likely to influence the messages that are transmitted. For example, stress alters the composition of microbial communities. Since these produce various neurotransmitters, such as GABA, as well as short-chain fatty acids, which act on the nervous system in various ways, their modification has a direct influence on the central nervous system. The addition of gut microbiota to mice that lack one is reflected, for example, in the tripling of serotonin levels in their blood plasma. The influence of microbiota on mood has been demonstrated in mice, in which oral administration of the pathogen *Campylobacter jejuni*, even in small doses, leads to anxious behavior [FOR 10]. In another field, we might cite the recent discovery of the role of digestive tract bacteria in circadian rhythms [DUB 15].

Studies have shown that manipulating digestive microbial communities could contribute to curing pathologies and perhaps even to influencing behavior. Trials of fecal microbiota transplantation are currently being carried out on patients suffering from Crohn's disease, with encouraging results and, notably, a reduction in symptoms [COL 14]. Today, the functional study of such microbial communities remains very difficult, due to their complexity, the multiplicity of interactions between all the components (microbial and tissue) and the high variability between the microbiota of different individuals [CON 12]. It is probably still hasty to hope to understand the mechanisms at play in each pathology, but curing disease by manipulating microbiota is a fascinating prospect. Intensive fundamental research is currently looking into model vertebrate organisms, such as mice and zebrafish (*Danio rerio*), which can be reared without microorganisms. They can therefore be inoculated with perfectly controlled communities, or with probiotic bacteria that may correct any imbalance. There can therefore be no doubt that major advances will be made in this field over the next few years.

7.2.2. *When the symbiont becomes an Achilles heel or a vaccine*

These last few decades have seen the resurgence of resistance to products such as antibiotics and insecticides usually used to fight pathogens and disease carriers. Alongside this continued chemical battle, research is exploring new strategies, including some that are connected to symbiosis. Alternative control strategies for disease-carrying animal populations could be based on targeting the symbiotic bacteria that they contain [SAS 13]. For example, new treatments against parasitic filarial nematode worms responsible for serious pathologies, such as river blindness and elephantiasis, are based on eliminating their obligate intracellular symbiont, *Wolbachia pipiensis*. Antibiotics such as doxycycline are used to eliminate the bacteria, sterilizing the female nematodes and ultimately killing all the individuals. This strategy avoids the use of compounds that are more toxic for the patient and produces excellent results [TAY 13].

Many insects besides nematodes harbor symbionts and strategies for controlling their populations by acting on the bacteria are beginning to be explored. For example, the tsetse fly, whose symbiosis with *Wigglesworthia* has been discussed, carries a trypanosome (*Trypanosoma brucei*) that causes sleeping sickness. Mosquitoes carry many diseases, including malaria, the Zika virus and chikungunya. As well as targeting bacterial symbionts through antibiotic treatments, another promising avenue in the fight against disease carriers involves using symbionts to make the insects less efficient at transmitting them. This is, for example, the aim of the international "Eliminate Dengue" (eliminatedengue.org) project, which aims to fight this viral disease that affects 50 million people and causes around 25,000 deaths per year. The mosquito that carries dengue, *Aedes aegypti*, is not associated with the bacterium *Wolbachia* in nature. By infecting the mosquito with some lineages of *Wolbachia*, researchers have discovered that its capacity to transmit the dengue virus diminishes considerably, because the virus's development is slowed down [MOR 09b]. The same seems to be true of other viruses, such as Zika and chikungunya. The project therefore aims to introduce mosquitoes harboring *Wolbachia* into natural populations of *Aedes aegypti*. Counting on this bacterium's ability to spread in populations by manipulating its hosts' reproduction, as described in section 3.3, researchers hope to lower the transmission of dengue, and perhaps of other viruses, in the most affected areas. In a way, introducing symbionts to make the insects unable to transmit diseases harks back to vaccinating the disease carriers themselves, rather than humans.

Yet more strategies exist, such as introducing lineages of symbionts that reduce the lifespan of insects. Through the destruction, introduction and manipulation of symbionts, we hope to find alternatives to the aggressive treatments for some human parasitoses and the mass spreading of insecticides to which disease-carrying insects gradually become resistant [RIC 12]. Alongside their value to human health, approaches using symbionts as Achilles heels or vaccines will also eventually be used to control crop pests.

7.2.3. *Symbiosis and plant production*

With more than 7.3 billion inhabitants in 2015, the population of Earth is six times higher today than it was in 1900. Fortunately, agricultural yields have increased in proportion, enabling us more or less to feed the world. We owe this to major evolutions in production methods since the first plants were domesticated around 10,000 years ago. Irrigation; varietal improvement, initially empirical and then scientific; crop rotation; fertilizers and pesticides have secured food supplies for a large number of humans. All the same, it is anticipated that food production will have to double over the next 40 years. Although the improvement of plants through genetic engineering is a possible avenue, the production of genetically modified organisms raises important ethical questions and is rejected by many consumers. Other avenues are being explored, including many directly linked to symbiosis.

Young plants already inoculated with specific mycorrhizal fungi are now commercially available. This method makes it possible to guarantee the presence of appropriate and efficient fungal partners, which can be selected depending on soil properties, for example. This enables things to be planted in poor soils, degraded by excessive amendments or sterilized under greenhouses, for example. It also makes it possible to produce economically relevant associations. Indeed, the French *Institut National pour la Recherche Agronomique* (INRA) proposes oak seedlings that have been mycorrhized by *Tuber melanosporum*, otherwise known as black truffle. Nowadays, 90% of truffles produced in France come from such plants. Endophytic fungi are another type of plant symbiont that is beginning to be used in agriculture. Chapter 3 described their protective properties against various stresses in nature (drought, salt, temperature and pathogens). However, the most remarkable discovery is that inoculating other plants with endophytes from plants living in stressful conditions transfers these same resistances to them and can even improve their productivity [VAR 99]. In recent years, important studies have therefore been carried out to develop plants equipped with endophytes that make them more resistant to stress and more economical when it comes to water. For example, field trials with treated maize and rice plants are being carried out in the United States and seem promising (Bioensure, Adaptive Symbiotic Technologies). It should be noted that until now this has involved approaches that are based not on genetic or chemical manipulation, but simply on the building of associations that exist or could exist in nature, and that these innovations can be used in sustainable, even organic agriculture. One final avenue of research aims to create Rhizobia, nitrogen-fixing bacteria present in the nodules of many plants, that are capable of nodulating grains through genetic engineering. Grains are not able to associate with nitrogen-fixing bacteria. Transferring the symbiosis found in legumes to cereals would enable the use of fertilizers in grain agriculture to be considerably reduced. This technology comes from the aforementioned synthetic biology, and although research in this field is still highly exploratory, the recent development of genome synthesis technologies makes it possible to begin to examine its practical feasibility [ROG 14].

7.2.4. *Avenues in animal productivity*

Just as for humans, the study of the microbiota associated with farmed animals is emerging as a topic whose impact seems promising. Since its origins at least 9,000 years ago, animal husbandry has benefited from breed improvement and, more recently, from industrialization, for better in terms of productivity and for worse in terms of animal well-being. Two current challenges are the optimization of alimentary efficiency, i.e. the mass of meat, eggs and milk produced per unit of food consumed, and the limitation of the use of antibiotics, which are now used not only against diseases but primarily as growth factors [ALL 13]. By comparing chickens that use food very efficiently to others that are less efficient, researchers have shown that their associated digestive communities are different, notably with lower microbial diversity in the most successful [STA 13]. Interestingly, lower diversity in the most efficient chickens (and therefore the fattest) recalls the same phenomenon previously mentioned in obese mice and humans. However, although obesity is viewed negatively in humans, the opposite is, of course, true in intensive farming, whose aim is precisely to obtain the largest quantity of biomass for a given food investment! From these results, the authors suggest manipulating the digestive microbiota of chickens to stabilize it and limit its diversity, in order to improve performances, and to achieve this by producing a mixture of microbes with which to inoculate the chicks when they hatch. Such avenues are harder to explore in larger animals with longer life cycles, but microbiota is a factor that will certainly be taken into account increasingly in future improvement strategies [MIG 15].

7.3. Some avenues in ecology

The knowledge of symbiotic associations acquired over the last few decades is beginning to filter into environmental sciences. This is particularly true in cases such as that of coral reefs. As well as being hotspots for marine biodiversity, they are significant for many human communities. These reefs protect coasts from erosion, directly feed more than 500 million people and generate economic activities such as tourism, which bring them income. Yet these coral reefs are in the front line in the face of global changes. The prolonged elevation of surface temperatures, even by one or two degrees, as well as other stresses, can cause a divorce between host and zooxanthellae, the latter being expelled. Massive bleaching episodes have recently affected up to 80% of corals in some areas. Although many are then recolonized by symbionts, these episodes can often lead to the death of 20% of corals, making the reefs fragile. Ocean acidification magnifies this phenomenon by making the precipitation of the calcareous skeleton of the corals more difficult. We can also add to this the damage caused by the often excessive human frequentation of some areas, destruction by predators, such as some invasive species of starfish, and natural disasters, such as the 2004 Indian Ocean tsunami. It is estimated that more than three-quarters of coral reefs are in danger. The success of many preservation and restoration programs is and will always be influenced by awareness of the needs of visible organisms, but also, of course, of their microbial symbionts. As such, the identification of some types of zooxanthellae that are particularly thermotolerant and therefore likely to maintain the association in a warmer ocean in corals in the Andaman Sea in the Indian Ocean opens up interesting avenues in conservation biology [HUM 15]. Alongside this example, which illustrates the importance of considering symbiotic relationships in habitat protection projects, other projects are beginning to be developed to evaluate the role of symbionts in the success of invasive species and to implement management strategies that integrate them [KOW 15].

The last few years have seen the emergence of microbiology as a major discipline in natural sciences. Now armed with adequate tools, research is revealing the immense diversity of microorganisms, their metabolic capacities and their role in ecosystems. By studying these communities of microorganisms, we can now start to understand their role in fields as varied as health, agriculture and the environment. It is becoming possible to envisage manipulating these communities to cure diseases, improve production methods, and maintain and restore ecosystems [ALI 15].

Conclusion and Perspectives

Following the work of pioneers at the end of the 19th and the beginning of the 20th Century, and the initial definition of symbiosis, the 1965 publication of the English version of Paul Buchner's synthesis, entitled "Endosymbiosis of animals with plant microorganisms", is often considered the founding act of a field of biology devoted to the study of symbiosis. Furthermore, it was in this period that the first conferences specifically devoted to the subject were held. Studies carried out prior to this had been focused on a limited number of models that could be kept in the laboratory and from which various hypotheses could be proposed and tested as to the nature and workings of symbiosis. Since Buchner, the increasingly detailed study of symbiosis has often benefited from methodological advances. Electron microscopy has brought about obvious progress in describing the structure of associations. The development of molecular methods facilitated the study of uncultivable systems and therefore the discovery of the abundance and frequency of symbiotic relationships in nature. The more we search, the more we discover. This has given us access to vital information concerning the functioning of symbioses, their distinctive features, those they have in common with other types of interaction, and even their evolution. We are currently experiencing a third wave of discoveries. The democratization of high-throughput analysis methods for genomes, transcriptomes, proteomes, lipidomes and metabolomes, capable of generating prodigious quantities of data, now makes it possible to study very complex systems and move from a reductionist towards a more holistic approach. This goes hand in hand with a significant paradigm shift: the awareness of the central importance of microbes on our planet, even in our own bodies [ALI 15, DUB 15]. Still, The best understood systems today are the simplest, in which the role of the partners can be delineated. However, understanding the interactions within communities as complex as the microbiota associated with the digestive tract or the skin in the same detail will probably be possible in the near future.

These revolutions substantially alter our views on biodiversity and life in general. Pluricellular organisms, of which we are an example, are merely the visible agents of an essentially microbial world. These organisms themselves are the habitat for bacteria. Microbial cells are often more numerous than hosts' cells, and hosts depend upon their microbial partners for many functions. Concepts that are central to our thinking in life sciences are being questioned. The concept of the holobiont, a composite entity that includes host and symbionts and on which natural selection acts, directly calls into question our classic concept of species. The dialogue established between partners can be compared with that between the various organs that make up an animal or a plant, and the mechanisms for transfer and acquisition of symbionts in each generation are part of the reproduction strategies. This directly calls into question the concept of the individual, as does a recent article entitled "A symbiotic view of life: we have never been individuals" [GIL 12]. To study the evolution of such systems, a vision of evolution reduced to a model based on mutation and selection and its representation by a dichotomous tree seems insufficient. Indeed, association is a force of biological innovation of the utmost importance that cannot be represented appropriately by this model.

Ultimately, the study of symbiosis calls into question even its own paradigms, since the amplitude of the discoveries leads it to redefine itself. The classic vision based on evaluation of cost and benefit seems less and less relevant as we discover that, in numerous cases, partners can live separately and the holobiont is the object of selection. But the holobiont itself is only one of the levels at which selection operates, since it can also function between the partners of the group. It is as if the holobiont were an ecosystem in which a variety of interconnected players are evolving. Without resolving all the ambiguities, definitions more centered on the acquisition of new functions and the emergence of a composite entity are more relevant. In a way, inapplicable definitions are often being replaced with ones that are applicable but more flexible. However, the latter are not the end of the story in any way and knowledge enhancement will probably produce new incarnations of this definition.

The study of symbioses is, by definition, interdisciplinary. Firstly in its objects, because it affects viruses, archaea, bacteria, protists, plants, fungi and animals. Then in its observation levels, because it examines the association from the molecule through to the ecosystem. Finally, in its tools, because it chooses from a wide range of available techniques and methods. Due to its interdisciplinarity, the study of symbiotic organisms has long been seen as a branch of any number of disciplines and often treated anecdotally. This is evident in most university courses, where its teaching occupies a negligible place, usually divided between various disciplines, such as physiology, evolution and ecology, which, furthermore, use different definitions. However, through recent discoveries that have made us aware of its importance in nature, symbiosis has begun to take its rightful place at the heart of life science disciplines. Associating is one of the major characteristics of the organisms that populate our planet, and understanding it requires an integrated approach, centered on the very notion of association.

Positioned at the center of biology, the study of symbioses offers fascinating and particularly promising avenues for research. This is true at the fundamental level, through the identification of new partners and new functions, but in other areas as well. In health, decrypting the links between microbiota and diseases could help us find cures in microflora, control pathogens through their symbionts or identify new therapeutic molecules. In agriculture, microbial cocktails could be used to improve the productivity, stress resistance and water use of plants, and symbionts should be integrated among the targets of biological pest control. In environmental sciences symbionts could be used to facilitate the maintenance or restoration of habitats. We are even beginning to see this concept of symbiosis emerging in the social sciences, to which it brings a new perspective that no longer necessarily sets nature and culture against one another [HIR 10]. A true domain of the future.

Bibliography

[ALI 15] ALIVISATOS A.P., BLASER M.J., BRODIE E.L. *et al.*, "MICROBIOME. A unified initiative to harness Earth's microbiomes", *Science*, vol. 350, pp. 507–508, 2015.

[ALL 13] ALLEN H.K., LEVINE U.Y., LOOFT T. *et al.*, "Treatment, promotion, commotion: antibiotic alternatives in food-producing animals", *Trends in Microbiology*, vol. 21, pp. 114–119, 2013.

[ARC 11] ARCHIE E.A., THEIS K.R., "Animal behaviour meets microbial ecology", *Animal Behaviour*, vol. 82, pp. 425–436, 2011.

[BAR 07] BARTON E.S., WHITE D.W., CATHELYN J.S. *et al.*, "Herpesvirus latency confers symbiotic protection from bacterial infection", *Nature*, vol. 447, pp. 326–329, 2007.

[BAU 14] BAUM D.A., BAUM B., "An inside-out origin for the eukaryotic cell", *BMC Biology*, vol. 12, p. 76, 2014.

[BHA 13] BHATTACHARYA D., PELLETREAU K.N., PRICE D.C. *et al.*, "Genome analysis of *Elysia chlorotica* egg DNA provides no evidence for horizontal gene transfer into the germ line of this kleptoplastic mollusc", *Molecular Biology and Evolution*, vol. 30, pp. 1843–1852, 2013.

[BOR 15] BORDENSTEIN S.R., THEIS K.R., "Host biology in light of the microbiome: ten principles of holobionts and hologenomes", *PLoS ONE*, vol. 13, p. e1002226, 2015.

[BRA 81] BRANDT K., "Uber Das Zusammenleben von Algen Und Tieren", *Biologisches Zentralblatt*, vol. 1, pp. 524–527, 1881.

[BRA 03] BRAENDLE C., MIURA T., BICKEL R. *et al.*, "Developmental origin and evolution of bacteriocytes in the aphid–*Buchnera* symbiosis", *PLoS Biology*, vol. 1, pp. 70–76, 2003.

[BRI 10] BRIGHT M., BUGHERESI S., "A complex journey: transmission of microbial symbionts", *Nature Reviews Microbiology*, vol. 8, pp. 218–230, 2010.

[BRO 09] BROWNLIE J.C., JOHNSON K.N., "Symbiont-mediated protection in insect hosts", *Trends in Microbiology*, vol. 17, pp. 348–354, 2009.

[BRU 14] BRUNE A., "Symbiotic digestion of lignocellulose in termite guts", *Nature Reviews Microbiology*, vol. 12, pp. 168–180, 2014.

[BUC 65] BUCHNER A., *Endosymbiosis of Animals with Plant Microorganisms*, Interscience, New York, 1965.

[CAN 05] CANFIELD D.E., "The early history of atmospheric oxygen: homage to Robert M. Garrels", *Annual Review of Earth and Planetary Sciences*, vol. 33, pp. 1–36, 2005.

[CLA 14] CLAY K., "Defensive symbiosis: a microbial perspective", *Functional Ecology*, vol. 28, pp. 293–298, 2014.

[CLE 64] CLEVELAND L.R., GRIMSTONE A.V., "The fine structure of the flagellate *Mixotricha paradoxa* and its associated microorganisms", *Proceedings of the Royal Society of London B*, vol. 159, pp. 668–686, 1964.

[COL 14] COLMAN R.J., RUBIN D.T., "Fecal microbiota transplantation as therapy for inflammatory bowel disease: a systematic review and meta-analysis", *Journal of Crohn's and Colitis*, vol. 8, pp. 1569–1581, 2014.

[COM 95] COMBES C., *Interactions Durables: Écologie et Évolution Du Parasitisme*, Masson, Paris, 1995.

[CON 10] CONSORTIUM TIAG, "Genome sequence of the pea aphid *Acyrthosiphon pisum*", *PLoS Biology*, vol. 8, no. e1000313, 2010.

[CON 12] CONSORTIUM THMP, "Structure, function and diversity of the healthy human microbiome", *Nature*, vol. 486, pp. 207–214, 2012.

[DAL 06] DALE C., MORAN N.A., "Molecular interactions between bacterial symbionts and their hosts", *Cell*, vol. 126, pp. 453–465, 2006.

[DAR 59] DARWIN C., *On the Origin of Species*, John Murray, London, 1859.

[DIL 02] DILLON R.J., VENNARD C.T., CHARNLEY A.K., "A note: gut bacteria produce components of a locust cohesion pheromone", *Journal of Applied Microbiology*, vol. 92, pp. 759–763, 2002.

[DOU 94] DOUGLAS A.E., *Symbiotic Interactions*, Oxford University Press, Oxford, 1994.

[DOU 98] DOUGLAS A.E., "Nutritional interactions in insect–microbial symbioses: aphids and their symbiotic bacteria *Buchnera*", *Annual Review of Entomology*, vol. 43, pp. 17–37, 1998.

[DOU 16] DOUGLAS A.E., WERREN J.H., "Holes in the hologenome: why host–microbe symbioses are not holobionts", *mBio*, vol. 7, p. e02099-15, 2016.

[DOV 01] DOVE S.G., HOEGH-GULDBERG O., RANGANATHAN S., "Major colour patterns of reef-building corals are due to a family of GFP-like proteins", *Coral Reefs*, vol. 19, pp. 197–204, 2001.

[DUB 08] DUBILIER N., BERGIN C., LOTT C., "Symbiotic diversity in marine animals: the art of harnessing chemosynthesis", *Nature Reviews Microbiology*, vol. 6, pp. 725–740, 2008.

[DUB 15] DUBILIER N., MCFALL-NGAI M., ZHAO L., "Microbiology: create a global microbiome effort", *Nature*, vol. 526, pp. 631–634, 2015.

[DUP 10] DUPERRON S., "The diversity of deep-sea mussels and their bacterial symbioses", in KIEL S. (ed.), *The Vent and Seep Biota*, Springer, pp. 137–167, 2010.

[DUP 13] DUPERRON S., GAUDRON S.M., RODRIGUES C.F. *et al.*, "An overview of chemosynthetic symbioses in bivalves from the North Atlantic and Mediterranean Sea", *Biogeosciences*, vol. 10, pp. 3241–3267, 2013.

[DUP 16] DUPERRON S., QUILES A., SZAFRANSKI K.M. *et al.*, "Estimating symbiont abundances and gill surface areas in specimens of the hydrothermal vent mussel *Bathymodiolus puteoserpentis* maintained in pressure vessels", *Frontiers in Marine Science*, vol. 3, p. 16, 2016.

[EAT 11] EATON C.J., COX M.P., SCOTT B., "What triggers grass endophytes to switch from mutualism to pathogenism?", *Plant Science*, vol. 180, pp. 190–195, 2011.

[END 90] ENDOW K., OHTA S., "Occurrence of bacteria in the primary oocytes of vesicomyid clam *Calyptogena soyoae*", *Marine Ecology Progress Series*, vol. 64, pp. 309–317, 1990.

[END 15] ENDERS G., ENDERS J., *Gut: the Inside Story of our Body's Most Underrated Organ*, Greystone Books, 2015.

[FAI 11] FAIRCLOUGH S.R., The cellular and molecular basis of multicellular development in the choanoflagellate *Salpingoeca rosetta*, eScholarship, PhD Thesis, University of California, 2011.

[FEL 14] FELLBAUM C.R., MENSAH J.A., CLOOS A.J. *et al.*, "Fungal nutrient allocation in common mycorrhizal networks is regulated by the carbon source strength of individual host plants", *New Phytologist*, vol. 203, pp. 646–656, 2014.

[FIN 08] FINLAY R.D., "Ecological aspects of mycorrhizal symbiosis: with special emphasis on the functional diversity of interactions involving the extraradical mycelium", *Journal of Experimental Botany*, vol. 59, pp. 1115–1126, 2008.

[FOR 10] FORSYTHE P., SUDO N., DINAN T. *et al.*, "Mood and gut feelings", *Brain, Behavior, and Immunity*, vol. 24, pp. 9–16, 2010.

[FRA 77] FRANK A.B., "Über die biologischen Verhältnisse des Thallus einiger Krustenflechten", *Beiträge zur Biologie der Pflanzen II*, vol. 2, pp. 123–200, 1877.

[FRA 04] FRANK A.B., "On the nutritional dependence of certain trees on root symbiosis with belowground fungi (an English translation of A.B. Frank's classic paper of 1885)", *Mycorrhiza*, vol. 15, pp. 267–275, 2004.

[GIB 10] GIBSON D.G., GLASS J.I., LARTIGUE C. *et al.*, "Creation of a bacterial cell controlled by a chemically synthesized genome", *Science*, vol. 329, pp. 52–56, 2010.

[GIL 12] GILBERT S.F., SAPP J., TAUBER A.I., "A symbiotic view of life: we have never been individuals", *Quarterly Review of Biology*, vol. 87, pp. 325–341, 2012.

[GOF 05] GOFFREDI S.K., ORPHAN V.J., ROUSE G.W. *et al.*, "Evolutionary innovation: a bone-eating marine symbiosis", *Environmental Microbiology*, vol. 7, pp. 1369–1378, 2005.

[GON 03] GONZÁLEZ J.E., MARKETON M.M., "Quorum sensing in nitrogen-fixing rhizobia", *Microbiology and Molecular Biology Reviews*, vol. 67, pp. 574–592, 2003.

[HAM 64] HAMILTON W.D., "The genetical evolution of social behaviour. I", *Journal of Theoretical Biology*, vol. 7, pp. 1–16, 1964.

[HAR 09] HARRIS R.N., BRUCKER R.M., WALKE J.B. *et al.*, "Skin microbes on frogs prevent morbidity and mortality caused by a lethal skin fungus", *ISME Journal*, vol. 3, pp. 818–824, 2009.

[HED 08] HEDGES L.M., BROWNLIE J.C., O'NEILL S.L. *et al.*, "*Wolbachia* and virus protection in insects", *Science*, vol. 322, p. 702, 2008.

[HER 87] HERRING P.J., "Systematic distribution of bioluminescence in living organisms", *Journal of Bioluminescence and Chemiluminescence*, vol. 1, pp. 147–163, 1987.

[HIR 10] HIRD M.J., "Coevolution, symbiosis and sociology", *Ecological Economics*, vol. 69, pp. 737–742, 2010.

[HON 00] HONEGGER R., "Simon Schwendener (1829–1919) and the dual hypothesis of lichens", *The Bryologist*, vol. 103, pp. 307–313, 2000.

[HON 07] HONGOH Y., SATO T., DOLAN M.F. *et al.*, "The motility symbiont of the termite gut flagellate *Caduceia versatilis* is a member of the "Synergistes" group", *Applied and Environmental Microbiology*, vol. 73, pp. 6270–6276, 2007.

[HON 08] HONGOH Y., SHARMA V.K., PRAKASH T. *et al.*, "Genome of an endosymbiont coupling N_2 fixation to cellulolysis within protist cells in termite gut", *Science*, vol. 322, pp. 1108–1109, 2008.

[HUM 15] HUME B.C.C., D'ANGELO C., SMITH E.G. *et al.*, "*Symbiodinium thermophilum* sp. nov., a thermotolerant symbiotic alga prevalent in corals of the world's hottest sea, the Persian/Arabian Gulf", *Scientific Reports*, vol. 5, no. 8562, 2015.

[HUN 99] HUNECK S., "The significance of lichens and their metabolites", *Naturwissenschaften*, vol. 86, pp. 559–570, 1999.

[JOH 11] JOHNSON M.D., "The acquisition of phototrophy: adaptive strategies of hosting endosymbionts and organelles", *Photosynthesis Research*, vol. 107, pp. 117–132, 2011.

[JOY 12] JOYE S.B., "Microbiology: a piece of the methane puzzle", *Nature*, vol. 491, pp. 538–539, 2012.

[KAD 05] KADAR E., BETTENCOURT R., COSTA V. *et al.*, "Experimentally induced endosymbiont loss and re-acquirement in the hydrothermal vent bivalve *Bathymodiolus azoricus*", *Journal of Experimental Marine Biology and Ecology*, vol. 318, pp. 99–110, 2005.

[KAM 14] KAMKE J., RINKE C., SCHWIENTEK P. *et al.*, "The candidate phylum Poribacteria by single-cell genomics: new insights into phylogeny, cell-compartmentation, eukaryote-like repeat proteins, and other genomic features", *PLoS ONE*, vol. 9, no. e87353, 2014.

[KAT 12] KATZ L.A., "Origin and diversification of eukaryotes", *Annual Review of Microbiology*, vol. 66, pp. 411–427, 2012.

[KAU 11] KAU A.L., AHERN P.P., GRIFFIN N.W. *et al.*, "Human nutrition, the gut microbiome and the immune system", *Nature*, vol. 474, pp. 327–336, 2011.

[KEE 08] KEELING P.J., ARCHIBALD J.M., "Organelle evolution: what's in a name?", *Current Biology*, vol. 18, pp. R345–R347, 2008.

[KIE 03] KIERS E.T., ROUSSEAU R.A., WEST S.A. *et al.*, "Host sanctions and the legume–rhizobium mutualism", *Nature*, vol. 425, pp. 78–81, 2003.

[KIE 06] KIERS E.T., VAN DER HEIJDEN M.G.A., "Mutualistic stability in the arbuscular mycorrhizal symbiosis: exploring hypotheses of evolutionary cooperation", *Ecology*, vol. 87, pp. 1627–1636, 2006.

[KIE 11] KIERS E.T., DUHAMEL M., BEESETTY Y. *et al.*, "Reciprocal rewards stabilize cooperation in the mycorrhizal symbiosis", *Science*, vol. 333, pp. 880–882, 2011.

[KLO 15] KLOSE J., POLZ M.F., WAGNER M. *et al.*, "Endosymbionts escape dead hydrothermal vent tubeworms to enrich the free-living population", *Proceedings of the National Academy of Sciences of the USA*, vol. 112, pp. 11300–11305, 2015.

[KNI 09] KNITTEL K., BOETIUS A., "Anaerobic oxidation of methane: progress with an unknown process", *Annual Review of Microbiology*, vol. 63, pp. 311–334, 2009.

[KOG 12] KOGA R., MENG X.-Y., TSUCHIDA T. *et al.*, "Cellular mechanism for selective vertical transmission of an obligate insect symbiont at the bacteriocyte–embryo interface", *Proceedings of the National Academy of Sciences of the USA*, vol. 109, pp. E1230–E1237, 2012.

[KOW 15] KOWALSKI K.P., BACON C., BICKFORD W. *et al.*, "Advancing the science of microbial symbiosis to support invasive species management: a case study on *Phragmites* in the Great Lakes", *Frontiers in Microbiology*, vol. 6, p. 95, 2015.

[KUW 07] KUWAHARA H., YOSHIDA T., TAKAKI Y. *et al.*, "Reduced genome of the thioautotrophic intracellular symbiont in a deep-sea clam, *Calyptogena okutanii*", *Current Biology*, vol. 17, pp. 881–886, 2007.

[LAN 07] LANYON C.V., RUSHTON S.P., O'DONNELL A.G. *et al.*, "Murine scent mark microbial communities are genetically determined", *FEMS Microbiology Ecology*, vol. 59, pp. 576–583, 2007.

[LAU 08] LAUER A., SIMON M.A., BANNING J.L. *et al.*, "Diversity of cutaneous bacteria with antifungal activity isolated from female four-toed salamanders", *ISME Journal*, vol. 2, pp. 145–157, 2008.

[LEC 07] LECHENE C.P., LUYTEN Y.A., MCMAHON G. *et al.*, "Quantitative imaging of nitrogen fixation by individual bacteria within animal cells", *Science*, vol. 317, pp. 1563–1566, 2007.

[LEL 12] LELIAERT F., SMITH D.R., MOREAU H. *et al.*, "Phylogeny and molecular evolution of the green algae", *Critical Reviews in Plant Sciences*, vol. 31, pp. 1–46, 2012.

[LEY 08] LEY R.E., LOZUPONE C.A., HAMADY M. *et al.*, "Worlds within worlds: evolution of the vertebrate gut microbiota", *Nature Reviews Microbiology*, vol. 6, pp. 776–788, 2008.

[LOP 14] LOPANIK N.B., "Chemical defensive symbioses in the marine environment", *Functional Ecology*, vol. 28, pp. 328–340, 2014.

[LUT 01] LUTZONI F., PAGEL M., REEB V., "Major fungal lineages are derived from lichen symbiotic ancestors", *Nature*, vol. 411, pp. 937–940, 2001.

[LUT 09] LUTZONI F., MIADLIKOWSKA J., "Lichens", *Current Biology*, vol. 19, pp. R502–R503, 2009.

[MAC 02] MACKIE R.I., "Mutualistic fermentative digestion in the gastrointestinal tract: diversity and evolution", *Integrative and Comparative Biology*, vol. 42, pp. 319–326, 2002.

[MAR 71] MARGULIS L., *Origin of Eukaryotic Cells*, Yale University Press, 1971.

[MÀR 07] MÀRQUEZ L.M., REDMAN R.S., RODRIGUEZ R.J. *et al.*, "A virus in a fungus in a plant: three-way symbiosis required for thermal tolerance", *Science*, vol. 315, pp. 513–515, 2007.

[MAZ 08] MAZMANIAN S.K., ROUND J.L., KASPER D.L., "A microbial symbiosis factor prevents intestinal inflammatory disease", *Nature*, vol. 453, pp. 620–625, 2008.

[MCC 12] MCCUTCHEON J.P., MORAN N.A., "Extreme genome reduction in symbiotic bacteria", *Nature Reviews Microbiology*, vol. 10, pp. 13–26, 2012.

[MCC 14] MCCUTCHEON J.P., KEELING P.J., "Endosymbiosis: protein targeting further erodes the organelle/symbiont distinction", *Current Biology*, vol. 24, pp. R654–R655, 2014.

[MCF 01] MCFADDEN G.I., "Primary and secondary endosymbiosis and the origin of plastids", *Journal of Phycology*, vol. 37, pp. 951–959, 2001.

[MCF 13] MCFALL-NGAI M.J., HADFIELD M.G., BOSCH T.C.G. *et al.*, "Animals in a bacterial world, a new imperative for the life sciences", *Proceedings of the National Academy of Sciences of the USA*, vol. 110, pp. 3229–3236, 2013.

[MCF 14a] MCFALL-NGAI M.J., "Divining the essence of symbiosis: insights from the squid-*Vibrio* model", *PLoS Biology*, vol. 12, no. e1001783, 2014.

[MCF 14b] MCFALL-NGAI M.J., "The importance of microbes in animal development: lessons from the squid-*Vibrio* symbiosis", *Annual Review of Microbiology*, vol. 68, pp. 177–194, 2014.

[MCF 15] MCFALL-NGAI M.J., "Giving microbes their due – animal life in a microbially dominant world", *Journal of Experimental Biology*, vol. 218, pp. 1968–1973, 2015.

[MER 09] MERCOT H., POINSOT D., "Infection by *Wolbachia*: from passengers to residents", *Comptes Rendus Biologies*, vol. 332, pp. 284–297, 2009.

[MIG 15] MIGNON-GRASTEAU S., NARCY A., RIDEAU N. *et al.*, "Impact of selection for digestive efficiency on microbiota composition in the chicken", *PLoS ONE*, vol. 10, no. e0135488, 2015.

[MIL 01] MILLER M.B., BASSLER B.L., "Quorum sensing in bacteria", *Annual Review of Microbiology*, vol. 55, pp. 165–199, 2001.

[MIL 12] MILUCKA J., FERDELMAN T.G., POLERECKY L. *et al.*, "Zero-valent sulphur is a key intermediate in marine methane oxidation", *Nature*, vol. 491, pp. 541–546, 2012.

[MIN 92] MINDELL D.P., "Phylogenetic consequences of symbioses: Eukarya and Eubacteria are not monophyletic taxa", *Biosystems*, vol. 27, pp. 53–62, 1992.

[MON 14] MONÉ Y., MONNIN D., KREMER N., "The oxidative environment: a mediator of interspecies communication that drives symbiosis evolution", *Proceedings of the Royal Society of London B*, vol. 281, no. 20133112, 2014.

[MOR 05] MORAN N.A., DAGNAN P.H., SANTOS S.R. *et al.*, "The players in a mutualistic symbiosis: insects, bacteria, viruses and virulence genes", *Proceedings of the National Academy of Sciences of the USA*, vol. 102, pp. 16919–16926, 2005.

[MOR 06] MORAN N.A., DUNBAR H.E., "Sexual acquisition of beneficial symbionts in aphids", *Proceedings of the National Academy of Sciences of the USA*, vol. 103, pp. 12803–12806, 2006.

[MOR 14] MORAN N.A., BENNETT G.M., "The tiniest tiny genomes", *Annual Review of Microbiology*, vol. 68, pp. 195–215, 2014.

[MOR 98] MOREIRA D., LOPEZ-GARCIA P., "Symbiosis between methanogenic archaea and delta-proteobacteria as the origin of eukaryotes: the syntrophic hypothesis", *Journal of Molecular Evolution*, vol. 47, pp. 517–530, 1998.

[MOR 09a] MOREIRA D., LOPEZ-GARCIA P., "Ten reasons to exclude viruses from the tree of life", *Nature Reviews Microbiology*, vol. 7, pp. 306–311, 2009.

[MOR 09b] MOREIRA L.A., ITUBE-ORMAETXE I., JEFFERY J.A. *et al.*, "A *Wolbachia* symbiont in *Aedes aegypti* limits infection with Dengue, Chikungunya, and *Plasmodium*", *Cell*, vol. 139, pp. 1268–1278, 2009.

[MOR 13] MORRIS B.E.L., HENNEBERGER R., HUBER H. *et al.*, "Microbial syntrophy: interaction for the common good", *FEMS Microbiology Reviews*, vol. 37, pp. 384–406, 2013.

[MOR 15] MORIYAMA M., NIKOH N., HOSOKAWA T. *et al.*, "Riboflavin provisioning underlies *Wolbachia*'s fitness contribution to its insect host", *mBio*, vol. 6, no. e01732-15, 2015.

[MOY 08] MOYA A., PERETO J., GIL R. *et al.*, "Learning how to live together: genomic insights into prokaryote–animal symbioses", *Nature Reviews Genetics*, vol. 9, pp. 218–229, 2008.

[NUS 06] NUSSBAUMER A.D., FISHER C.R., BRIGHT M., "Horizontal endosymbiont transmission in hydrothermal vent tubeworms", *Nature*, vol. 441, pp. 345–348, 2006.

[NYH 04] NYHOLM S.V., MCFALL-NGAI M.J., "The winnowing: establishing the squid–*Vibrio* symbiosis", *Nature Reviews Microbiology*, vol. 2, pp. 632–642, 2004.

[OHK 08] OHKUMA M., "Symbioses of flagellates and prokaryotes in the gut of lower termites", *Trends in Microbiology*, vol. 16, pp. 345–352, 2008.

[OLI 08] OLIVER K.M., CAMPOS J., MORAN N.A. *et al.*, "Population dynamics of defensive symbionts in aphids", *Proceedings of the Royal Society of London B*, vol. 275, pp. 293–299, 2008.

[OLI 14] OLIVER K.M., SMITH A.H., RUSSELL J.A., "Defensive symbiosis in the real world – advancing ecological studies of heritable, protective bacteria in aphids and beyond", *Functional Ecology*, vol. 28, pp. 341–355, 2014.

[PEE 98] PEEK A.S., FELDMAN R.A., LUTZ R.A. *et al.*, "Cospeciation of chemoautotrophic bacteria and deep sea clams", *Proceedings of the National Academy of Sciences of the USA*, vol. 95, pp. 9962–9966, 1998.

[PÉR 06] PÉREZ-BROCAL V., GIL R., RAMOS S. *et al.*, "A small microbial genome: the end of a long symbiotic relationship?", *Science*, vol. 314, pp. 312–313, 2006.

[PLA 11] PLAISANCE L., CALEY M.J., BRAINARD R.E. *et al.*, "The diversity of coral reefs: what are we missing?", *PLoS ONE*, vol. 6, no. e25026, 2011.

[POR 13] PORCAR M., LATORRE A., MOYA A., "What symbionts teach us about modularity", *Frontiers in Bioengineering and Biotechnology*, vol. 1, p. 14, 2013.

[RAI 00] RAI A.N., SODERBACK E., BERGMAN B., "Tansley Review No. 116. *Cyanobacterium*–plant symbioses", *New Phytologist*, vol. 147, pp. 449–481, 2000.

[RAV 13] RAVEN J.A., "Rubisco: still the most abundant protein of Earth?", *New Phytologist*, vol. 198, pp. 1–3, 2013.

[RED 99] REDMAN R.S., FREEMAN S., CLIFTON D.R. *et al.*, "Biochemical analysis of plant protection afforded by a nonpathogenic endophytic mutant of *Colletotrichum magna*", *Plant Physiology*, vol. 119, pp. 795–804, 1999.

[REM 16] REMIGI P., ZHU J., YOUNG J.P.W. *et al.*, "Symbiosis within symbiosis: evolving nitrogen-fixing legume symbionts", *Trends in Microbiology*, vol. 24, pp. 63–75, 2016.

[RIC 12] RICCI I., VALZANO M., ULISSI U. *et al.*, "Symbiotic control of mosquito borne disease", *Pathogens and Global Health*, vol. 106, pp. 380–385, 2012.

[RID 13] RIDAURA V.K., FAITH J.J., REY F.E. *et al.*, "Gut microbiota from twins discordant for obesity modulate metabolism in mice", *Science*, vol. 341, no. 1241214, 2013.

[ROD 08] RODRIGUEZ R.J., HENSON J., VAN VOLKENBURGH E. *et al.*, "Stress tolerance in plants via habitat-adapted symbiosis", *ISME Journal*, vol. 2, pp. 404–416, 2008.

[ROD 09] RODRIGUEZ R.J., WHITE J.F., ARNOLD A.E. *et al.*, "Fungal endophytes: diversity and functional roles", *New Phytologist*, vol. 182, pp. 314–330, 2009.

[ROG 14] ROGERS C., OLDROYD G.E.D., "Synthetic biology approaches to engineering the nitrogen symbiosis in cereals", *Journal of Experimental Botany*, vol. 65, pp. 1939–1946, 2014.

[ROO 11] ROOSSINCK M.J., "The good viruses: viral mutualistic symbioses", *Nature Reviews Microbiology*, vol. 9, pp. 99–108, 2011.

[RUM 11] RUMPHO M.E., PELLETREAU K.N., MOUSTAFA A. *et al.*, "The making of a photosynthetic animal", *Journal of Experimental Biology*, vol. 214, pp. 303–311, 2011.

[SAB 09] SABREE Z.L., KAMBHAMPATI S., MORAN N.A., "Nitrogen recycling and nutritional provisioning by *Blattabacterium*, the cockroach endosymbiont", *Proceedings of the National Academy of Sciences of the USA*, vol. 106, pp. 19521–19526, 2009.

[SAC 10] SACHS J.L., EHINGER M.O., SIMMS E.L., "Origins of cheating and loss of symbiosis in wild *Bradyrhizobium*", *Journal of Evolutionary Biology*, vol. 23, pp. 1075–1089, 2010.

[SAC 11] SACHS J.L., SKOPHAMMER R.G., REGUS J.U., "Evolutionary transitions in bacterial symbiosis", *Proceedings of the National Academy of Sciences of the USA*, vol. 108, pp. 10800–10807, 2011.

[SAP 94] SAPP J., *Evolution by Association*, Oxford University Press, 1994.

[SAS 13] SASSERA D., EPIS S., PAJORO M. *et al.*, "Microbial symbiosis and the control of vector-borne pathogens in tsetse flies, human lice, and triatomine bugs", *Pathogens and Global Health*, vol. 107, pp. 285–292, 2013.

[SCH 04] SCHMIDT H., HENSEL M., "Pathogenicity islands in bacterial pathogenesis", *Clinical Microbiology Reviews*, vol. 17, pp. 14–56, 2004.

[SEL 15] SELOSSE M.-A., STRULLU-DERRIEN C., "Origins of the terrestrial flora: a symbiosis with fungi?", in MAUREL M.-C., GRANDCOLAS P. (eds), *BIO Web Conferences*, vol. 4, p. 9, 2015.

[SHI 00] SHIGENOBU S., WATANABE H., HATTORI M. *et al.*, "Genome sequence of the endocellular bacterial symbiont of aphids *Buchnera* sp. APS", *Nature*, vol. 407, pp. 81–86, 2000.

[SHI 11] SHIGENOBU S., WILSON A.C.C., "Genomic revelations of a mutualism: the pea aphid and its obligate bacterial symbiont", *Cellular and Molecular Life Sciences*, vol. 68, pp. 1297–1309, 2011.

[SMI 87] SMITH D.C., DOUGLAS A.E., *The Biology of Symbiosis*, Edward Arnold, London, 1987.

[SNY 13] SNYDER A.K., RIO R.V.M., "Interwoven biology of the tsetse holobiont", *Journal of Bacteriology*, vol. 195, pp. 4322–4330, 2013.

[STA 13] STANLEY D., GEIER M.S., DENMAN S.E. *et al.*, "Identification of chicken intestinal microbiota correlated with the efficiency of energy extraction from feed", *Veterinary Microbiology*, vol. 164, pp. 85–92, 2013.

[STI 05] STINGL U., RADEK R., YANG H. *et al.*, "Endomicrobia: cytoplasmic symbionts of termite gut protozoa form a separate phylum of prokaryotes", *Applied and Environmental Microbiology*, vol. 71, pp. 1473–1479, 2005.

[TAI 02] TAIZ L., ZIEGER E., *Plant Physiology*, 4th ed., Sinauer Associates, Inc., 2002.

[TAM 02] TAMAS I., KLASSON L., CANBÄCK B. *et al.*, "50 million years of genomic stasis in endosymbiotic bacteria", *Science*, vol. 296, pp. 2376–2379, 2002.

[TAY 13] TAYLOR M.J., VORONIN D., JOHNSTON K.L. *et al.*, "*Wolbachia* filarial interactions", *Cellular Microbiology*, vol. 15, pp. 520–526, 2013.

[THE 12] THEIS K.R., SCHMIDT T.M., HOLEKAMP K.E., "Evidence for a bacterial mechanism for group-specific social odors among hyenas", *Scientific Reports*, vol. 2, no. 615, 2012.

[THO 12] THONG-ON A., SUZUKI K., NODA S. *et al.*, "Isolation and characterization of anaerobic bacteria for symbiotic recycling of uric acid nitrogen in the gut of various termites", *Microbes and environments / JSME*, vol. 27, pp. 186–192, 2012.

[TIT 96] TITLYANOV E.A., TITLYANOVA T.V., LELETKIN V.A. *et al.*, "Degradation of zooxanthellae and regulation of their density in hermatypic corals", *Marine Ecology Progress Series*, vol. 139, pp. 167–178, 1996.

[TSU 04] TSUCHIDA T., KOGA R., FUKATSU T., "Host plant specialization governed by facultative symbiont", *Science*, vol. 303, p. 1989, 2004.

[TUR 07] TURNBAUGH P.J., LEY R.E., HAMADY M. *et al.*, "The Human Microbiome Project", *Nature*, vol. 449, pp. 804–810, 2007.

[TUR 09] TURNBAUGH P.J., HAMADY M., YATSUNENKO T. *et al.*, "A core gut microbiome in obese and lean twins", *Nature*, vol. 457, pp. 480–484, 2009.

[VAN 75] VAN BENEDEN P.-J., *Animal Parasites and Messmates*, D. Appleton & Company, New York, 1875.

[VAR 99] VARMA A., VERMA S., SUDHA, *et al.*, "*Piriformospora indica*, a cultivable plant-growth-promoting root endophyte", *Applied and Environmental Microbiology*, vol. 65, pp. 2741–2744, 1999.

[WAR 07] WARNECKE F., LUGINBÜHL P., IVANOVA N. *et al.*, "Metagenomic and functional analysis of hindgut microbiota of a wood-feeding higher termite", *Nature*, vol. 450, pp. 560–565, 2007.

[WEB 12] WEBSTER N.S., TAYLOR M.W., "Marine sponges and their microbial symbionts: love and other relationships", *Environmental Microbiology*, vol. 14, pp. 335–346, 2012.

[WEN 03] WENZEL M., RADEK R., BRUGEROLLE G. *et al.*, "Identification of the ectosymbiotic bacteria of *Mixotricha paradoxa* involved in movement symbiosis", *European Journal of Protistology*, vol. 39, pp. 11–23, 2003.

[WER 97] WERREN J.H., "Biology of *Wolbachia*", *Annual Review of Entomology*, vol. 42, pp. 587–609, 1997.

[WER 02] WERNEGREEN J.J., "Genome evolution in bacterial endosymbionts of insects", *Nature Reviews Genetics*, vol. 3, pp. 850–861, 2002.

[WER 08] WERREN J.H., BALDO L., CLARK M.E., "*Wolbachia*: master manipulators of invertebrate biology", *Nature Reviews Microbiology*, vol. 6, pp. 741–751, 2008.

[WID 10] WIDDER E.A., "Bioluminescence in the ocean: origins of biological, chemical, and ecological diversity", *Science*, vol. 328, pp. 704–708, 2010.

[WIE 00] WIER A., MARGULIS L., "The wonderful lives of Joseph Leidy (1823–1891)", *International Microbiology*, vol. 3, pp. 55–58, 2000.

[WOE 77] WOESE C.R., FOX G.E., "Phylogenetic structure of the prokaryotic domain: the primary kingdoms", *Proceedings of the National Academy of Sciences of the USA*, vol. 74, pp. 5088–5090, 1977.

[ZOO 15] ZOOK D., "Symbiosis – evolution's co-author", in GONTIER N. (ed.), *Reticulate Evolution*, Springer, pp. 41–80, 2015.

Index

M, N, O

P, Q, R

S, T

V, W, Z

Printed in the United States
By Bookmasters